CREATING MIND

How the Brain Works

CREATING MIND

How the Brain Works

John E. Dowling

W. W. NORTON & COMPANY
NEW YORK LONDON

Printed in the United States of America
First Edition

The text of this book is composed in Dante with the display set in Bernhard Modern.
Composition by JoAnn Schambier
Manufacturing by the Maple-Vail Book Manufacturing Group
Book design by Chris Welch

Figure 3C: Dowling, J. E. Synaptic organization of the frog retina: An electron microscopic analysis comparing the retinas of frogs and primates. *Proceedings of the Royal Society, B* 170 (1968): 205–228. Used by permission of the author and The Royal Society, London.

Figure 43: Block, J. R., and Yuker, H. E. *Can You Believe Your Eyes?: over 250 illusions and other visual oddities*. New York: Gardner Press, Inc., 1989. Copyright 1989 by Gardner Press, Inc. Courtesy of T. E. Parks and Peter Thompson, University of York, England.

Figure 44: Boring, E. G. Size Constancy in a Picture. *American Journal of Psychology* 77 (1964). Copyright 1964 by University of Illinois Press.

Library of Congress Cataloging-in-Publication Data
Dowling, John E.
Creating mind : how the brain works / John E. Dowling.
p. cm.
Includes bibliographical references and index.
ISBN 0-393-02746-5
1. Brain. 2. Neurosciences. I. Title.
QP376.D695 1998
612.8'2--dc21 98-9365
CIP

W. W. Norton & Company, Inc., 500 Fifth Avenue, New York, N.Y. 10110
http://www.wwnorton.com

W. W. Norton & Company Ltd., 10 Coptic Street, London WC1A 1PU

1 2 3 4 5 6 7 8 9 0

FOR JESSICA AND NICHOLAS

Contents

Preface ix

CHAPTER 1 The Uniqueness of the Brain 3

CHAPTER 2 Brain Signals 19

CHAPTER 3 Drugs and the Brain 41

CHAPTER 4 Creepy, Crawly Nervous Systems 61

CHAPTER 5 Brain Architecture 81

CHAPTER 6 Vision: Window to the Mind 101

CHAPTER 7 Development and Brain 123

CHAPTER 8 Language, Memory, and the Human Brain 143

CHAPTER 9 The Emotional Brain 161

CHAPTER 10 The Conscious Brain 177

Glossary 191

Further Reading 205

Index 207

Preface

What makes us human and unique among all creatures is our brain. Perception, consciousness, memory, learning, language, and intelligence all originate in and depend on the brain. The brain provides us with wondrous things, from mathematical theories to symphonies, from automobiles and airplanes to trips to the moon. But when it goes awry, we are undone.

Over the past century, our understanding of the brain has raced forward, yet those who study the brain are still scratching the surface, so to speak. What is the mind, after all, and how does it relate to brain function? Most neuroscientists believe the mind originates in brain function, but at the moment no one can define adequately what we mean by mind. Even "consciousness" is an elusive subject, though philosophers and others endlessly talk about what it means.

As a neuroscientist, I am forever peppered with questions about the brain and brain function. This is especially true for friends who know about the exciting discoveries in the brain sciences, yet witness the consequences of mental illness, aging, or brain injury and want to know more about new drug therapies for treating these problems. No field of medicine is untouched by the advances in the brain sciences, especially as we have come to realize how much the course of a disease and even its outcome, perhaps, can be affected by brain function and mental state. Sound body / sound mind is a two-way street; each is profoundly affected by the other.

This book is intended to answer many of the questions about neuroscience I am often asked. At the same time, I hope to convey to the general reader the essence and vitality of the field—the progress we are making in understanding how brains work—and to describe some of our strategies for studying brain function. Whenever possible, I try to relate topics to something relevant such as a disease or a consequence of brain function. Much wonderful work in the field is ignored—to keep the book manageable and, I hope, an interesting read.

The first five chapters provide the nuts and bolts necessary for an up-to-date understanding of the brain. The remainder of the book dips into aspects of brain function—vision, perception, language, memory, emotion, and consciousness—seemingly more relevant to how the brain creates mind. But if an in-depth understanding of these topics is to be gained, the nuts and bolts of brain function must be sorted out. Some readers may wish to start with Chapter 6 or another chapter of particular interest and return to the earlier chapters later for more details of how the brain works. A glossary is provided to help along the way with unfamiliar terms or concepts, but in almost every case further explanation is provided elsewhere in the book.

Chapter 1 describes the general organization of the brain. What are the cells like that are found in the brain? How do they differ from cells elsewhere in the body? Chapter 2 discusses how brain cells receive, process, and transmit information. Neural signals travel along cells electrically but between cells chemically. How do cells accomplish this? How do brain cells generate electrical signals, and how do chemical substances pass information to adjacent cells? Chapter 3 discusses in more detail how brain cells talk to one another and the changes that occur in brain cells when they are contacted by other cells. The chemical substances used to communicate signals in the brain are described and drugs that alter chemical transmission and cause profound alterations in brain function discussed.

Chapter 4 describes how invertebrates—animals without backbones—have been invaluable for elucidating neural mechanisms. Animals from the sea, such as squids, horseshoe crabs, and sea slugs, have been particularly useful, and examples of important findings from these animals are presented. Chapter 5 describes the architecture of the human brain—the various parts of the brain and what roles they play. How do the brains of frogs and fishes differ from our brains? The dominant role of the cerebral cortex in mammals and, especially, man is emphasized.

In Chapter 6, I explore the visual system in depth, from the light-sensitive molecules found in the photoreceptor cells to current ideas about visual perception. We know more about the visual system than about any other brain system; it provides a wealth of clues about brain function. Chapter 7 deals with the development of the brain. How do embryonic brain cells find their way to their targets? How does environment affect the developing brain? The fascinating topics of language, memory, and learning are dealt with in Chapter 8, along with the question of how we discover new things about the human brain. The neurology clinic has long provided instructive examples of patients with specific brain lesions. Today, brain imaging techniques promise a wealth of new information about the human brain.

Chapter 9 turns to matters we associate more with mind—emotions, personality, and rationality. What regions of the brain are involved in emotional behaviors, and what happens when these areas are disrupted? Finally, Chapter 10 discusses consciousness. What do we mean by the word, and what can we say about consciousness from a neurobiological perspective? Throughout the book, I use as examples instances where brain function is compromised by injury or disease. Such examples are not presented simply as curiosities. Rather, these alterations in brain function cast light on normalcy.

Howard Boyer guided the early stages of the book, providing superb suggestions and masterful editing of the first drafts. Laura Simonds Southworth executed the illustrations expertly and beautifully. Richard Mixter enthusiastically saw the book through to publication. Barbara Whitesides made many perceptive suggestions, and last, but certainly not least, Stephanie Levinson provided all the administrative help that a project like this entails.

CREATING MIND

How the Brain Works

CHAPTER 1

The Uniqueness of the Brain

Bob Jones was sixty two years old when he retired as chief executive officer of a small company. He had always been an effective administrator and was dedicated to his wife and family. Two years earlier, he had become uncharacteristically short-tempered, but this change was attributed by everyone to stress. He also failed on occasion to run promised errands, but his wife assumed he simply did not want to do them. Although retired as CEO, he continued to work for the company.

Over the next few months, Mr. Jones became increasingly forgetful, and finally he went to see his physician, who reassured him that he was just getting a little older. His irritability also increased, eventually to the point that nothing seemed right, and his wife took him back to the physician. No evidence of brain disease was evident, and the physician thought Mr. Jones was experiencing depression. An antidepressant drug was prescribed, along with psychotherapy, and both appeared to help. Mr. Jones seemed better for about a year.

Over the second year, however, Mr. Jones's memory deteriorated significantly, and he began experiencing attention loss and an inability to learn new things. He no longer could work, and a brain scan at this point indicated a significant shrinkage of his brain. A diagnosis of Alzheimer's disease was made.

Mr. Jones's mental abilities now began to deteriorate dramatically. On several occasions, he made a cup of coffee when his wife was out and forgot to turn the stove off. If he left the house, he promptly became lost, and eventually he became lost in his own home. He became so confused about right and left that he could not put on his clothes without help. Difficulty in orienting his arms and head in space eventually made it impossible for him to eat without assistance. His wife cared for him over this time but felt increasingly that it was not her husband she was caring for but a stranger. He had lost virtually all the traits that had made him a unique individual.

—Adapted from David L. Rosenham and Martin E. P. Seligman, *Abnormal Psychology: Casebook and Study Guide* (New York: Norton, 1995)

The human brain weighs no more than 3½ pounds, only about 2–3 percent of our total body weight, but its importance cannot be overstated. It oversees virtually everything we do and makes us what we are. When the brain deteriorates, as happened to Mr. Jones, not only are individuals unable to carry out even simple tasks such as eating, they also lose their uniqueness and individuality.

We are aware of many activities the brain controls; walking, talking, laughing, and thinking are just a few of them. The brain initiates these activities and also controls and regulates them. But, as we go about our daily lives, we are unaware of many other aspects of brain function: the regulation of internal organs, including the heart and vascular system, lungs and respiratory system, and gut and digestive system. The brain also coordinates and integrates movements employing mechanisms that we don't notice, such as the use of abundant sensory information from muscles, tendons, and joints. The extent of muscle contraction is signaled to the brain, yet we are quite unaware ordinarily of the state of our individual muscles.

Of most interest, and most mysterious, are mental functions referred to as "mind." Feelings, emotions, awareness, understanding, and creativity are well-known aspects of mind. Are they created in and by the brain? The consensus today among neuroscientists and philosophers is that mind is an emergent property of brain function. That is, what we refer to as mind is a natural consequence of complex and higher neural processing. Clearly brain injury or disease can severely compromise the mind, as happened to Mr. Jones. At the very least, then, mind depends on intact and healthy brain function.

Do animals have minds? The answer to this question depends mainly on one's definition of mind. Certainly cats, dogs, and monkeys can express emotion, show some understanding, and even apply creative approaches to simple problems, but no animal approaches humans in richness of mind. Is there something unique about human brains relative to those of other organisms? Not that we know. So how do we explain our extraordinary mental abilities?

Is it that human brains represent a higher evolutionary level than the brains of other animals? This is likely so, in part because the cellular mechanisms underlying human brain function appear identical to those operating in other animals, even those that have very elementary nervous systems and exhibit virtually no aspects of mind.

The simple view I espouse is that the human brain is qualitatively similar to the brains of other animals, but quantitatively different. That is, the human brain has more nerve cells than do the brains of other primates, our closest relatives, but also, what is probably more important, the cerebral cortex of the human brain, the seat of higher neural function—perception, memory, language, and intelligence—is far more developed than is the cerebral cortex of any other vertebrate (see Figure 1). *And because of the added neural cells and cortical development in the human brain, new facets of mind emerge.*

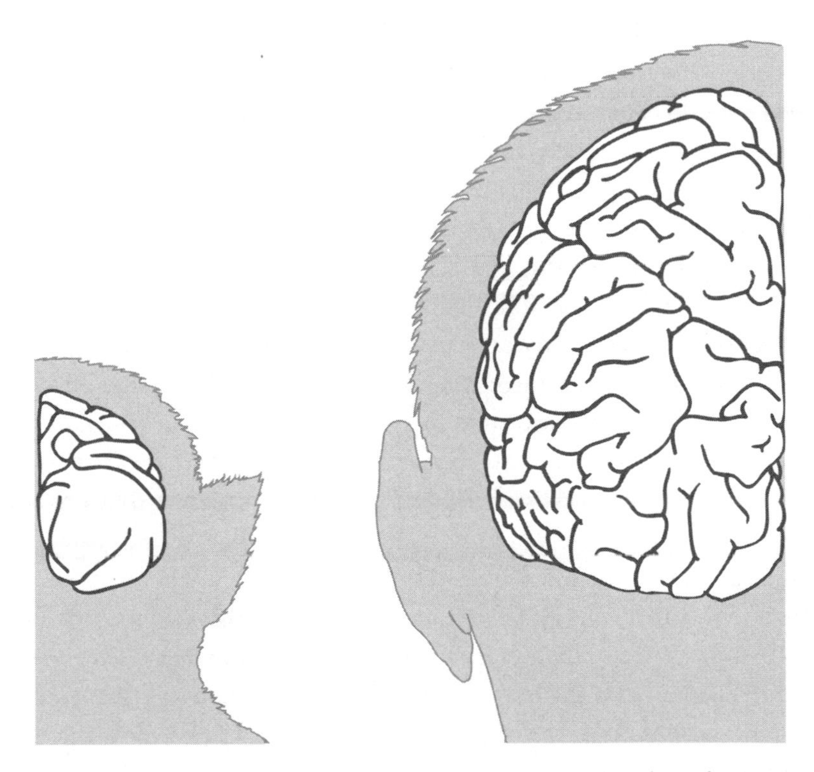

Figure 1 The brain of an adult rhesus monkey compared to that of a human. Not only is the human brain very much larger, its surface is thrown into many more folds, thus increasing the surface (cortical) area of the brain substantially.

An understanding of the brain requires knowing its structure, function, and chemistry. Once we do so, can we truly understand the mind? Can the brain understand itself? No one knows; there is much to learn.

To lay the groundwork for this quest, I'll describe the brain's elements and their structure, how the elements communicate with one another, their special features, and the consequences of these features. This will bring us to a general notion of how the brain is organized.

Cells of the Brain: Neurons

The brain, like other organs, is made up of discrete units or cells. Two classes of cell make up the brain: *neurons*, or nerve cells, whose business is to receive, integrate, and transmit information; and *glia* (derived from the Greek word for glue), which are supporting cells. Glial cells do things like help maintain the neurons and the brain's environment. They regulate the levels of substances needed or used by neurons in the spaces between the cells. They also provide a structural framework for neurons (especially during development), and they insulate the neurons to make them conduct electrical signals more effectively. But the key cells for understanding how the brain works are the neurons, and the brain contains billions of them.

We don't know exactly how many neurons are in the human brain, but the best estimates suggest between 100 billion and 1,000 billion—more cells than there are stars in the Milky Way. Neurons, like glial cells, are also elaborate and have numerous extensions or branches that may extend long distances. For example, a neuron that controls muscles in the foot has a branch, called an *axon*, that extends down the length of the leg to the foot, a distance of about 3 feet (or a meter). The body of this cell resides in the lower part of the spinal cord and is less than 0.1 mm in diameter. To put the difference between the size of the cell body and the length of its axon in perspective, consider how long the axon would be if the cell body were 6 inches wide—the axon would extend almost a mile!

The branches of neurons allow them to contact one another in the brain in complex and intricate ways. Typically, neurons make one hundred to ten thousand connections with other neurons, and one type of neuron (the cerebellar Purkinje cell) makes as many as 100,000 connections.

Neurons have many branches, and so most of the brain consists of neuronal branches. Two kinds of neuronal branches are distinguished anatomically—*dendrites* and *axons* (*see Figure 2*). Dendrites are like the branches of a tree, relatively thick as they emerge from the cell but dividing often and becoming thinner at each branch point. Many dendrites usually extend from each neuron. Axons, in contrast, are thinner at their point of origin on a neuron and remain constant in diameter along most of their length. Neurons usually have just one axon that

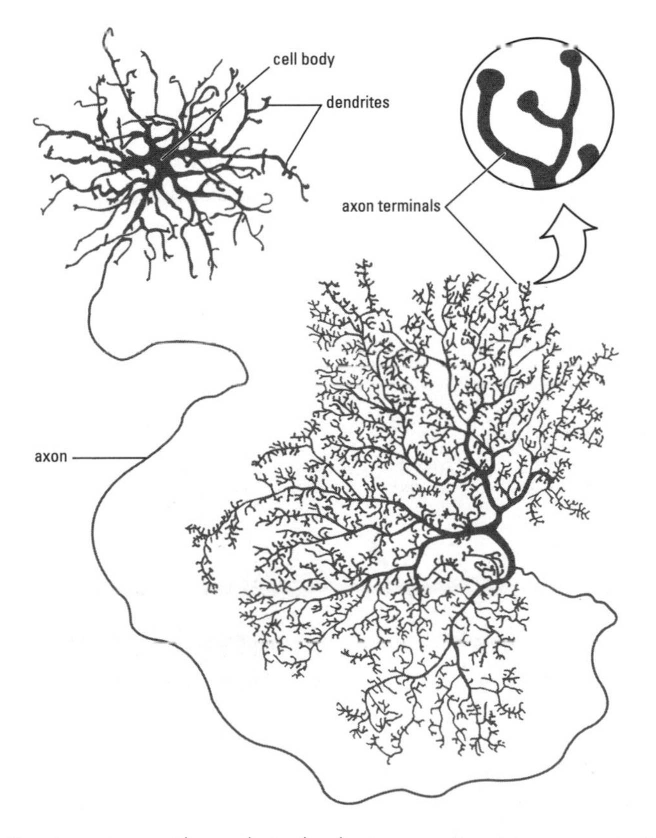

Figure 2 A neuron with a relatively short axon. Input to a neuron is usually onto the dendrites or cell body, whereas the cell's output occurs via the axon terminals. Each tiny branch in both the dendritic and axonal terminal arbors is likely to be a site of synaptic contact.

branches profusely as it terminates. Input to nerve cells is usually via the dendrites; the axon carries the cell's output.

Figure 2 depicts a neuron with a short axon found in the retina of the eye. Each tiny branch in both the dendritic tree and axon terminal complex probably represents a point of functional contact with another cell. And it is likely that a number of contact points were missed by the scientist who drew the cell. The functional contacts between neurons, called *synapses*, operate mainly chemically. That is, at synapses neurons release specific chemical substances that diffuse to an adjacent neuron

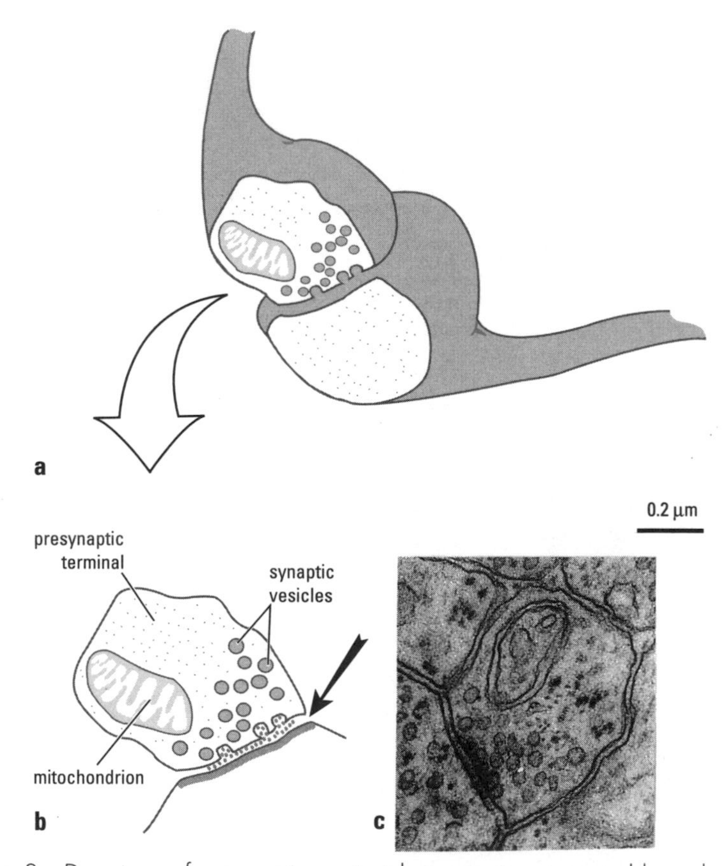

Figure 3 Drawings of a synaptic contact between two neuronal branches (above and below to the left) and an electron micrograph of a synapse (below right). The synaptic vesicles contain substances that are released when the synapse is active. The vesicles attach to the membrane and release their contents into the small space between the two branches (arrow). The substances released can excite, inhibit, or modulate activity in the contacted branch.

contacted by that synapse. This chemical may excite, inhibit, or modulate the contacted neuron.

Synapses have a characteristic structure that can be readily visualized with the high resolving power of an electron microscope. *Figure 3a* is a drawing of a synapse (essentially one of the terminal branches in *Figure 2*). Below to the right is an electron micrograph of a synapse and drawing of it on the left. What you are viewing in the electron micrograph is a slice through the middle of the axon terminal that is shown in the drawing above. A prominent feature of axon terminals is the presence of *synaptic vesicles* within the terminal: tiny vesicles that store the chemicals released at the synapse. The vesicles cluster at the synapse, fuse to the membrane when synapses are active, and release their contents into the small space between the two cells (*arrow in Figure 3*). The chemicals spread across the space and interact with specific molecules (proteins) present within the membrane of the contacted cell. The molecules, when activated by the synaptic chemical, initiate mechanisms that alter the cells.

Neuroscientists focus much attention on synapses because of the general agreement that synaptic interactions between neurons explain much of what the brain does. For example, most drugs that affect states of behavior, such as cocaine, LSD, Prozac, and even Valium, do so by modifying synaptic activity (see Chapter 3, Drugs and the Brain). Furthermore, affective mental disorders, such as schizophrenia, depression, and anxiety, appear to result from impaired synaptic mechanisms in the brain.

How Special Are Neurons?

Neurons employ the same cellular mechanisms as do other cells in the body. Each neuron has a nucleus containing *nucleic acid* (DNA) that specifies the proteins made by the cell. Neurons possess *ribosomes*, structures responsible for assembling proteins, and *mitochondria*, which supply energy-rich molecules that power the cells. Neurons also contain tubules and filaments found in virtually all cells; these are involved in the movement of substances throughout the cell, and they help maintain the complex structure of a neuron.

It is clear from this that the neurons' biochemical mechanisms are similar to those used by all cells. Yet neurons differ from other cells in the body in two significant ways which have important medical consequences: First, brain cells are not replaced or replenished and second, brain cells constantly require oxygen. Once neurons have matured during embryonic development, they can never divide again. This is quite different from most cells in the body that divide and produce new cells in response to injury or disease. A cut on the finger soon heals as cells divide and fill in the injury. The same is true for cells in most organs of the body.

When brain cells are lost because of injury or disease, they are not replaced. The brain of a one-year-old human contains as many cells as it will ever have, and throughout life neurons are lost via normal aging processes. In other organs, dead cells are quickly replaced, but in the brain they are not. And the number of brain cells lost is surprisingly high—maybe as many as 200,000 per day. This estimate comes from the finding that, with age, at least 5–10 percent of brain tissue is lost. If you assume a loss of 7 percent of the brain cells in eighty to one hundred years, with, say, 100 billion cells at the outset, about 200,000 cells are lost per day. With age, the brain also loses neuronal branches and the neurons shrink in size.

If neurons cannot divide once they have matured, how can a brain tumor grow? In adults, most if not all brain tumors are glial cell tumors. Glial cells, unlike neurons, can divide in the adult brain, and when glial cell division becomes uncontrolled, a tumor or cancer can result. Only in children do brain tumors arise from neurons, and, fortunately, these are quite rare.

Because the brain contains so many neurons, most of us can get through life without losing so many cells that we become mentally debilitated. Eventually, though, brain cell loss with age does catch up with us, and eventually mental deterioration takes place in virtually everyone. It's a mystery why some individuals maintain keen mental abilities much longer than do others; indeed, it may be brain cell loss that determines human life span. If we could eliminate heart attacks, cancer, and other fatal diseases, we still might not extend the life span of humans, because brain cells cannot divide and replace themselves.

Although the average life expectancy for humans has increased by about 50 percent in this century, from about fifty years in 1900 to seventy-five years today, the maximum number of years that humans live has not increased since ancient times. *Figure 4* shows the trends in human longevity from antiquity to the present day. Medical advances and improved housing and sanitation have greatly increased the numbers of people who live to age sixty, from about 20 percent of the population to nearly 80 percent. But only a very small percent of humans live to one hundred years of age, and this percentage has not changed much since antiquity.

Alzheimer's disease is marked by excessive brain cell loss. As many as four million Americans probably suffer from the disease, and it is estimated that as many as fourteen million may be affected by the year 2040. Those who suffer from it may have a decline in their mental abilities in their late fifties or early sixties, beginning with deficits in recent memory and progressing to a loss of virtually all higher mental functions. The symptoms displayed by Bob Jones described at the beginning of this chapter are typical. Confusion and forgetfulness, followed by sharply declining motor abilities and even loss of speech, are com-

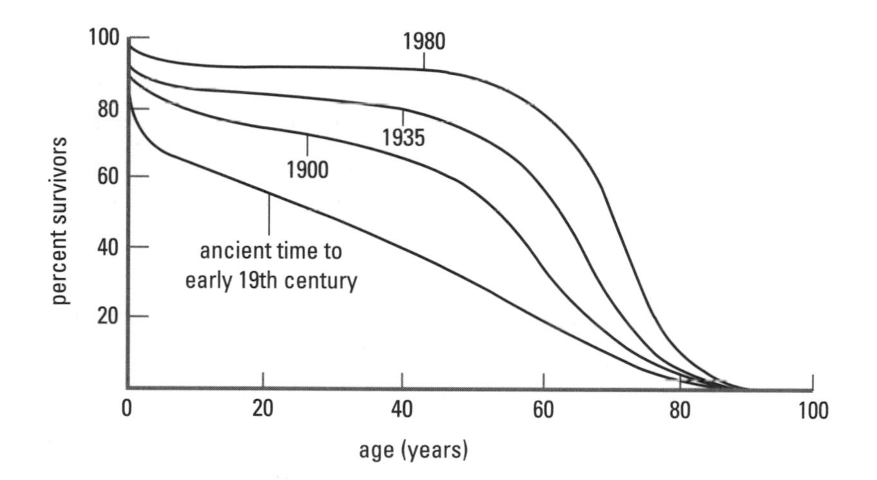

Figure 4 A graph showing the changes in life expectancy and life span of humans from ancient times to 1980. Although life expectancy has changed dramatically from an average of thirty-five years two hundred years ago to about seventy-five years today, the life span of humans has not increased significantly since antiquity. That is, only a very small percentage of humans live beyond one hundred years of age.

mon symptoms. Whether the brain cell loss in Alzheimer's disease is an acceleration of normal brain cell loss or the result of a disease is still unclear, although most neurologists believe the latter is so. By the age of sixty-five, about 10 percent of the population show Alzheimer-like symptoms, and by age eighty-five as much as half the population may have the symptoms.

Severe brain injury or trauma also leads to excessive brain cell loss, and these individuals often display Alzheimer-like symptoms. People prone to brain injury, like prizefighters, have these symptoms and are sometimes referred to as "punch-drunk." Brain cell death may not be evenly distributed throughout the brain, which leads to different symptoms for people with Alzheimer's disease, brain injury, or normal aging. The former heavyweight boxing champion Muhammad Ali is said to have Parkinson's disease, a degenerative brain disease affecting parts of the brain's motor control system. He walks sluggishly, has difficulty initiating movements, and displays an expressionless face, all classic symptoms of Parkinson's disease. It is likely that his disease is linked to the pounding his head took during his many years as a prizefighter.

Alzheimer's disease is probably not a single entity but several diseases. For example, early-onset Alzheimer's disease is clearly inherited in a number of instances, but in most cases (about 90 percent) the disease is probably not directly inherited. It may be that certain individuals have a genetic predisposition for Alzheimer's disease, but what the factors are that cause some individuals such as Bob Jones to develop the disease in their sixties and others never or very much later in life is not understood.

The second critical difference between neurons and cells from other tissues is that brain cells constantly require oxygen. When deprived of it, neurons die within a few minutes. Other cells can survive without oxygen and even function without oxygen (anaerobically) for a while. During a 100-yard race, a runner's leg-muscle cells use up the available oxygen after only about 30 yards. The muscles keep functioning without oxygen by chemically breaking down sugar and producing energy-rich molecules by a fermentation-like process. After the race, as oxygen is restored to the muscles, the cells break down these sugar fragments and restore the muscle to its resting state. The oxygen required to repay the oxygen debt is provided after the race when the runner breathes hard for several minutes. If your muscles are sore after heavy exertion, the reason is partly the excessive buildup of the breakdown products of sugar, especially lactic acid, during the anaerobic phase of muscle activity.

Neurons, however, cannot survive anaerobically, even for a short time. After a person has a heart attack or suffocates, oxygen flow to the body's tissues is shut off and the brain will quickly die. If oxygen is restored in just a few minutes, the brain can survive. It is not uncommon for a patient suffering a severe heart attack to be left permanently brain-dead after a short period of oxygen deprivation, whereas today, with effective life-support systems, other organs like the heart and kidneys survive and recover completely.

The brain's need for oxygen is so acute that when a part of the brain is active, blood flow rapidly increases in that region. This can be measured by imaging techniques like positron emission tomography (PET) scanning and functional magnetic resonance imaging (fMRI), which enable neuroscientists and physicians to probe the brains of awake subjects, even when they are involved in specific behaviors (*see* Chapter 8 and *Figure 56*). These powerful techniques provide neuroscientists and physicians with a wealth of information about the brain's multiplicity of roles and with critical clinical information about brain disease.

When a person suffers a stroke, blood flow to the brain is interrupted, and the region losing its blood supply usually dies because of the lack of oxygen and loss of nutrients. Although an immediate deficit is evident in most stroke patients, everyone who has observed a surviving stroke patient knows that some recovery occurs, and it can continue for months, long after the first acute changes, such as swelling of the brain, have subsided. The recovery is sometimes virtually complete. The same can happen after a severe brain injury; remarkable recovery sometimes takes place.

If brain cells die after a person has a stroke or injury, and if they are not replaced, how does one recover, especially over the long term? The answer seems to be that remaining brain cells are used; that is, nearby cells can take over for the damaged or dead brain cells, which enables at least a partial recovery. How extensive the recovery is depends on the damage and on the region damaged. When some parts of the brain are damaged, no recovery whatsoever is seen, but many parts are more forgiving or plastic. How other neurons can take over for lost or damaged cells is not well understood, but the sprouting of new processes by the remaining neurons may happen along with the formation of new synapses. The mechanisms for this are probably similar to those that occur in the developing brain, discussed in Chapter 7.

Brain Organization

The brain is far from homogeneous. It consists of many parts, each concerned with a separate facet of neural function. Furthermore, each part of the brain has quite distinct neuron shapes. These differences in structure presumably relate to the role of the cells and that part of the brain. *Figure 5* shows three cells, two from one region of the brain, the other from a second brain region. The enormous dissimilarities in cell structures are readily apparent. We don't yet understand why neurons have the shapes they do, and finding out the reasons is an enticing challenge for neuroscientists.

The distinct types of neurons are easily recognizable between individuals and even between species. That is, a pyramidal cell (*Figure 5a*) in the cortex of the human brain looks similar to a pyramidal cell in the rabbit's cortex. Likewise, a Purkinje cell (named after its discoverer, Jan Purkinje, a Czech whom many regard as the founder of histology, the study of tissues) in the monkey looks like a cat's Purkinje cell (*Figure 5c*). But pyramidal cells are easily distinguished from Purkinje cells and would never be confused, regardless of the animal brain in which the cell is observed.

How many different kinds of cells are there in a particular brain region? In any one part of the brain, there are usually only a few major cell types. The cerebellum has five major cell types, the retina five types, and the cerebral cortex just two. All cells of a specific type look quite similar to other cells of that type and play a similar functional role. But in many cases anatomists have divided the major cell types into subtypes, of which ten to twenty can exist in a given region. Phys-

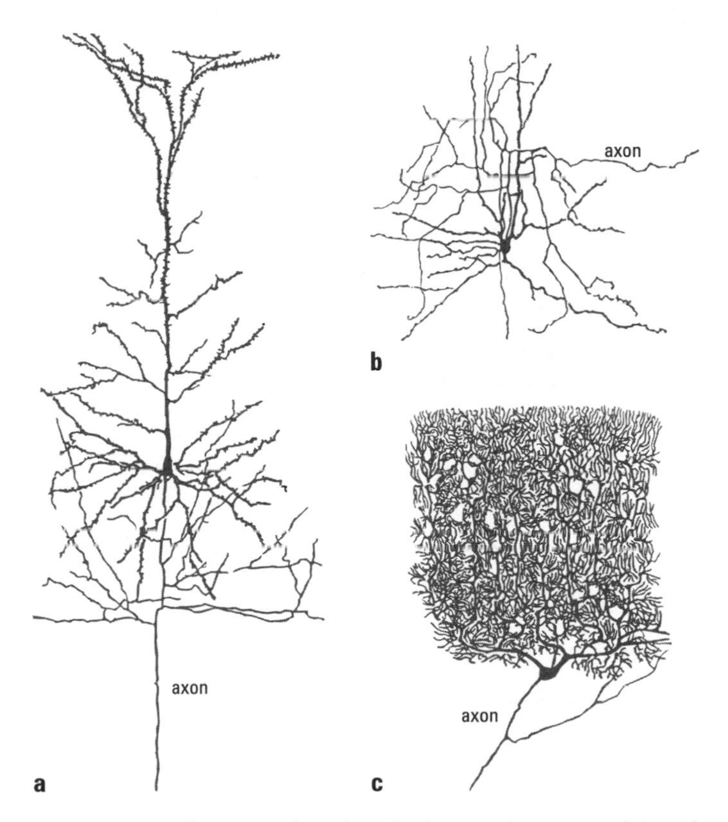

Figure 5 Three types of neurons found in the brain. The pyramidal and stellate cells (*a* and *b* respectively) are present in the cerebral cortex, whereas the Purkinje cell (*c*) is found in the cerebellum. These drawings illustrate the enormous diversity of neuronal structure found in the brain.

iologically the subtypes often respond in subtly different ways, so anatomical differences do have functional meaning.

It is convenient to divide the various cell types in the brain into two major classes: long-axon cells and short-axon cells. Long-axon cells carry information from one part of the brain to another; short-axon cells are confined to a single part of the brain (*Figure 5b*). Short-axon cells participate in local interactions between neurons and for this reason are often called *association neurons*; they are deeply involved in integrating and processing information. The greatest of the early brain histologists, Santiago Ramón y Cajal of Spain, who carried out his brain studies from the early 1880s to the 1930s, pointed out that brains of more highly developed animals contain relatively more short-axon cells than long-axon cells. This is indicative of the role that short-axon cells play in complex brain processing.

Throughout most parts of the brain, neurons are clustered together to form structures called nuclei. A brain nucleus usually carries out a specific neural task; for instance, a nucleus in one part of the brain regulates heart rate, while another nucleus in the same part of the brain controls respiration. Nuclei in another part of the brain regulate body temperature, hunger, or thirst, whereas nuclei in yet other parts of the brain are involved in the initiation of movements, or in the transmission of sensory information from lower brain centers to higher ones.

The general organization of a nucleus is illustrated in *Figure 6*. Information arrives at a nucleus via the axons of long-axon cells. The terminals of these cells synapse onto the dendrites of long-axon and short-axon cells. The branches of the short-axon cells are typically

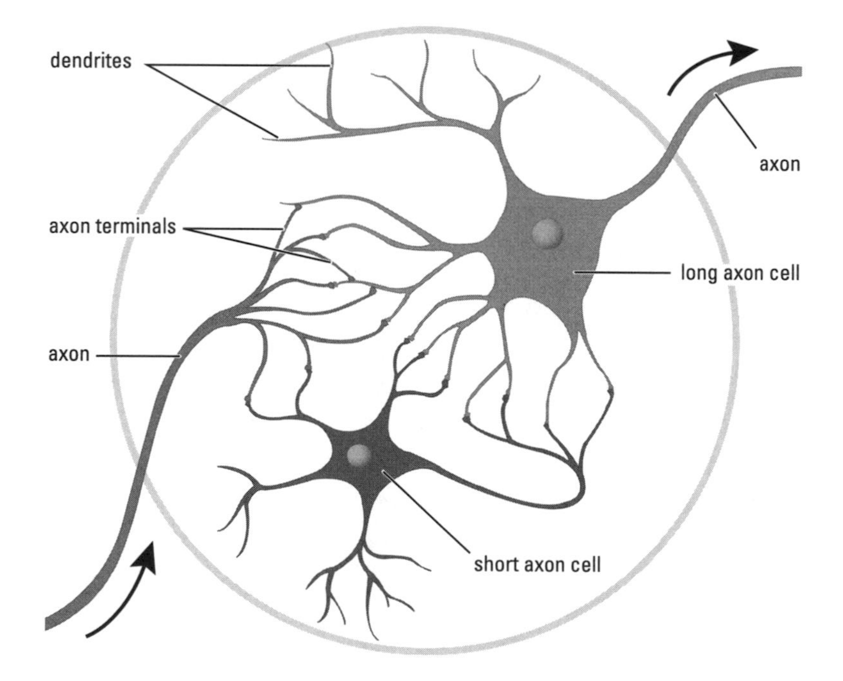

Figure 6 A schematic drawing of a brain nucleus. Information enters and exits (arrows) the nucleus via the axons of the larger, long-axon cells. Smaller short-axon cells mediate synaptic interactions within the nucleus—their branches are confined to the nucleus.

confined to the nucleus and make all of their synapses within the nucleus, on long-axon cells or other short-axon cells. Information leaves the nucleus via the axons of long-axon cells.

In some regions of the brain, for reasons not understood, neurons are arranged in continuous layers rather than in discrete nuclei, (*Figure 7*). But the same general principles of organization persist. That is, the short-axon neurons are restricted to a relatively local area, whereas long-axon neurons carry information from one brain area to another. The cerebral cortex is organized this way. However, it is also the case that the cortex is divided into areas that carry out specific tasks. For example, one area, V4 (V for visual), found in that part of the cortex concerned with vision primarily processes color information, whereas an adjacent area, V5, is concerned with analyzing moving stimuli. Such cortical areas are equivalent in functional terms to a brain nucleus, but are not as well defined anatomically. In other words, it is often difficult to decide where one cortical area ends and another begins.

An individual nucleus or a specific area of the brain is thus charged with a certain neural task, but the tasks are not equivalent. Some nuclei and brain areas process or regulate a basic neural function, but other nuclei and brain areas tackle higher or more subtle aspects of processing. For example, severe damage to the primary visual area, V1, brings about complete and permanent loss of visual perception

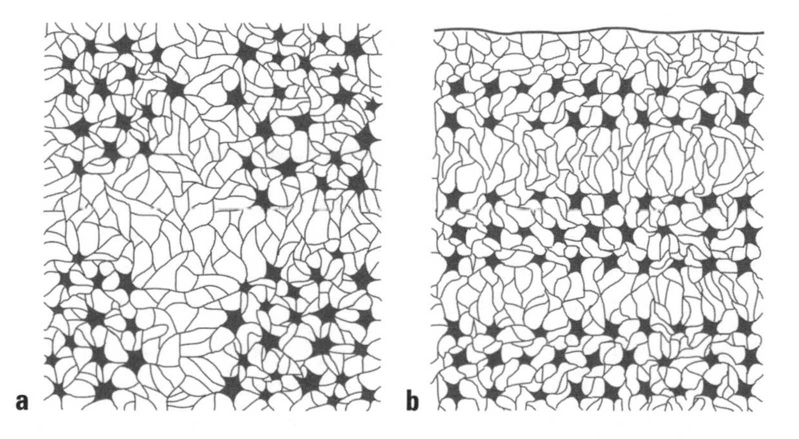

Figure 7 The common arrangements of neurons in the brain. In many regions, neurons are clustered in nuclei as shown on the left; in other regions, neurons are arranged in layers as on the right.

related to that part of the visual field served by that portion of the cortex. Yet similar damage to area V4 may lead only to a loss of color perception; other aspects of vision are maintained. And damage to yet higher visual areas might lead only to difficulty in recognizing or remembering specific objects.

The consequence of this hierarchical organization is that tiny lesions in some parts of the brain can have devastating effects while large lesions elsewhere can have minimal effects. Furthermore, after damage to certain areas or nuclei, little if any recovery happens; but damage to other areas may cause little if any long-term deficit. The bullet that killed John F. Kennedy entered a part of the brain where many nuclei regulate vital bodily functions such as respiration and heart rate. This region is also, from an evolutionary standpoint, an older part of the brain and more inflexible. Scant recovery happens when this part of the brain is damaged. A similar bullet wound in the neocortex—a more recently evolved brain region concerned with higher mental processing—might have caused slight permanent damage, and President Kennedy could conceivably still be alive today.

Neuroscientists are trying to discover why recovery from injury to some areas is much more complete than that from injury to other areas. A generalization emerging is that more recently evolved parts of the brain, those concerned with higher neural functioning, are more flexible or plastic than are older parts of the brain. This plasticity is reflected in the ability of a part of the brain to reorganize itself after damage and to recover function. Stated in cellular terms, neurons in some parts of the brain are more capable of extending new branches and forming new synapses than are neurons in other parts of the brain.

The brain's structure, then, is very heterogeneous, made up of many distinct areas concerned with different neural functions. Furthermore, the neurons found in various brain regions are often anatomically distinct, and these structural differences are thought to relate to the role of a specific brain area. The next chapter explores how individual neurons work. Surprisingly, perhaps, all neurons function in basically similar ways. Understanding how one neuron works tells us pretty much how all neurons work. This is not to say that all brain neurons function identically, but that the general principles outlined in the following chapter hold for all neurons.

CHAPTER 2

Brain Signals

The kidneys cleanse the blood, ridding the body of waste products. In a single day, the blood is washed about ten times. Kidney failure results in a condition called uremia because of a dramatic increase in urea levels in the blood. Urea is the chief waste product of nitrogen metabolism in the body and is the major dissolved substance in urine. In acute kidney failure, urea levels can increase in the blood by as much as twentyfold to reach levels of up to 2 grams per liter—a truly prodigious amount!

Patients with uremia exhibit a variety of symptoms, many of which reflect upsets in brain, neural, and muscle mechanisms. Severe headaches occur early, along with muscle twitching in the limbs. Visual disturbances are frequently reported as well. Convulsions and coma occur in the later stages of uremia with death ensuing, usually because of fibrillation of the heart—uncoordinated and uncontrolled twitching of the heart muscle.

What causes these neurological and cardiac catastrophes in uremia? Not the large increases in blood urea levels as one might think, but much more modest increases in one positively charged ion—potassium (K^+). In kidney failure, K^+ ion levels go up from about 0.3 gram per liter of blood to about 0.8 g/l, but this change is sufficient to cause death and the neurological symptoms described above. Death by lethal injection, now a favored method of execution, results from raising blood K^+ levels by the direct injection of K^+ into the bloodstream. Understanding the key role of K^+ (and other ions) in generating and regulating the electrical activity of the brain, nerve, and muscle cells has been one of the triumphs of twentieth-century biology.

During the last half of this century, neuroscientists have made astonishing progress in understanding how individual neurons function. The structures of neurons and synapses have been elucidated in great detail. We understand how single nerve cells are excited by sensory stimuli such as light or sound, or synaptic input, and how neurons carry and code information. The challenge today is to understand how aggregates of nerve cells interact to produce meaningful patterns of neural activity and control ultimately the behavior of an animal. How does the activity of and interactions between neurons give rise to seemingly simple behaviors such as walking, swimming, and moving one's eyes? Much present-day research in neuroscience addresses these questions. Investigators are trying to relate a behavior of an animal to the activity of specific neurons or groups of neurons in its brain. To understand how neurons operate, we need to look beneath the surface and see how single neurons process and carry information and how groups of neurons communicate with one another.

Neurons use chemical and electrical signals to transmit and carry information. Communication between neurons (at synapses) is mainly chemical, whereas individual neurons usually code and carry information electrically. In this chapter I describe how the electrical and chemical signals are generated, how they encode information, and what happens when the signals are interrupted. To start, I provide a short primer on electricity.

Electricity and the Brain

Electrical charge is an intrinsic property of all matter. It comes in two polarities, positive and negative. Charges of the same polarity repel; charges of the opposite polarity attract. Two of the fundamental particles that make up atoms, *electrons* and *protons*, have charge. Electrons possess negative charge, whereas protons possess positive charge. The electricity used at home consists of electrons moving through metal wires. Electrical *current* is a measure of how many electrons move through a conducting medium, such as metal wire, during a period of time.

Atoms ordinarily have equal numbers of electrons and protons and are, therefore, electrically neutral. But atoms can gain or lose electrons

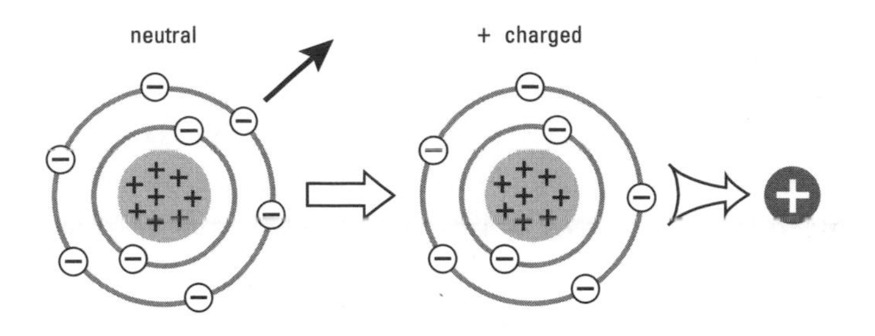

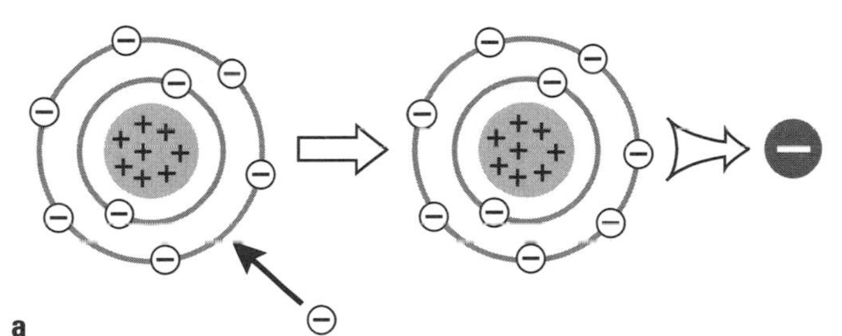

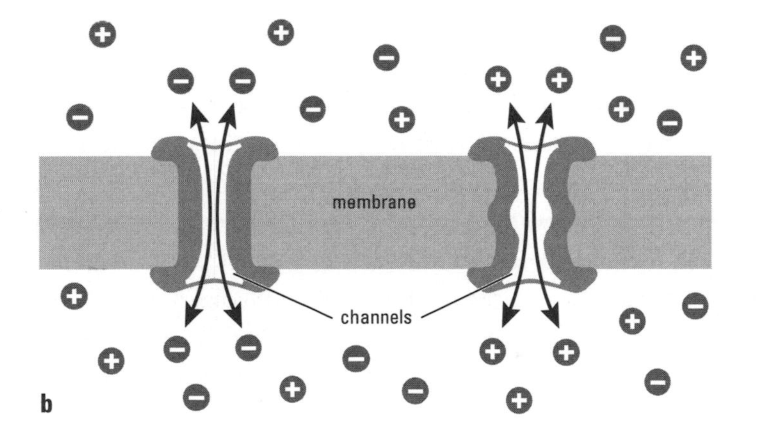

Figure 8 *a*: Electrically-neutral atoms become positively charged ions (top) or negatively charged ions (below) by losing or gaining an electron. A positively charged ion has more positively charged protons in the nucleus than negatively charged electrons surrounding the nucleus. A negatively charged ion has more electrons than protons.

b: A cell membrane is impermeable to ions. However, protein channels that span the membrane allow ions to flow from one side of the membrane to the other. Some channels allow only negatively charged ions to cross the membrane (left), others only positively charged ions (right).

and become charged, either negatively or positively (*Figure 8a*). They can have extra electrons or fewer electrons. Charged atoms are called *ions*; they generate the electrical signals in the brain when they move in and out of cells. Most ions have one extra positive or negative charge, but some ions important for neural function have two positive charges.

A cell's outer membrane does not readily allow ions to cross, but proteins in the membrane can form *channels* that allow and regulate the flow of ions across the membrane (*Figure 8b*). These channels can often discriminate among ions of different charge (such as sodium, a positively charged ion, versus chloride, a negatively charged ion) and even among different ions carrying the same charge (such as sodium versus potassium). A few channels are open all the time, but most channels allow ions to cross the membrane only in response to a stimulus, such as a chemical released at a synapse.

To keep ions flowing through the channels when they are open, cells maintain different concentrations of ions on either side of the membrane. They accomplish this with ion pumps: specific membrane proteins that actively transport ions from one side of the cell membrane to the other. The pumps require energy and are generally always operating, moving a few ions at a time across the membrane. When a channel opens, ions move through it in an attempt to equalize ion concentrations on both sides of the membrane. When ions cross the membrane, they alter the charge across the membrane. Membrane *voltage* (or *potential*) is a measure of the charge difference across a membrane. Voltages of 10–100 millivolts (0.01–0.1 volt) typically occur across cell membranes when a cell generates a signal. These are tiny voltages compared to the 110 volts powering our home electrical appliances.

Cell Resting Potentials

In addition to the electrical signals generated when nerve cells are stimulated, a resting voltage (also called the *resting potential*) exists across the outer membrane of all cells. In nerve cells, this resting voltage is about 70 millivolts (0.07 volt). The inside of the nerve cell is negatively charged relative to the outside; there are more negatively charged ions inside the cell than positively charged ions, whereas outside the cell there is an excess of positively charged ions. This charge imbalance results in the resting voltage that is measured between the inside and outside of the cell, across the cell's membrane. Understanding how resting potentials come about can help us understand how other electrical signals are produced.

Two factors underlie the resting potential of cells. First, ions are unevenly distributed on the two sides of a cell's membrane as a result of the action of the ion pumps present in the membrane. Four types of ions are principally involved: sodium (Na^+), potassium (K^+), chloride (Cl^-), and small organic molecules that have an excess negative charge and act like ions (A^-). Potassium ions (K^+) and the small negatively charged organic molecules (A^-) are predominantly inside cells, whereas Na^+ and Cl^- are mainly outside cells. Second, the membrane is differentially permeable to these ions. It is most leaky to K^+, least to A^-, and has low to moderate permeability to Na^+ and Cl^-. This means that nerve cell membranes in the resting state have channels that allow K^+ to pass through them, but no channels that allow A^- to pass through and few channels that allow Na^+ and Cl^- to pass through.

How a difference in ion concentrations between the inside and outside of a cell, coupled with a differential permeability of the cell's membrane to these ions, results in a resting voltage can be illustrated by a simple model of a cell that involves just K^+ and A^- (*Figure 9*). As in a real cell, we make the concentrations of K^+ and A^- high inside our model cell and low outside. Furthermore, we make the membrane permeable to K^+ but not to A^-. In other words, the channels in the membrane allow K^+ to pass through but not A^-. Now what happens? Both K^+ and A^- would like to equilibrate—that is, to move from an area of high concentration (inside the cell) to one of lower concentration (outside the cell). Potassium can pass across the membrane, but A^- cannot. Some K^+ leaves the cell, and for every K^+ that exits the cell, an extra negative charge is left inside and an extra positive charge is added to the outside of the cell. A voltage difference thus develops between inside and out; the inside of the cell is more negative (has excess negative charges) than outside the cell (which has excess positive charges).

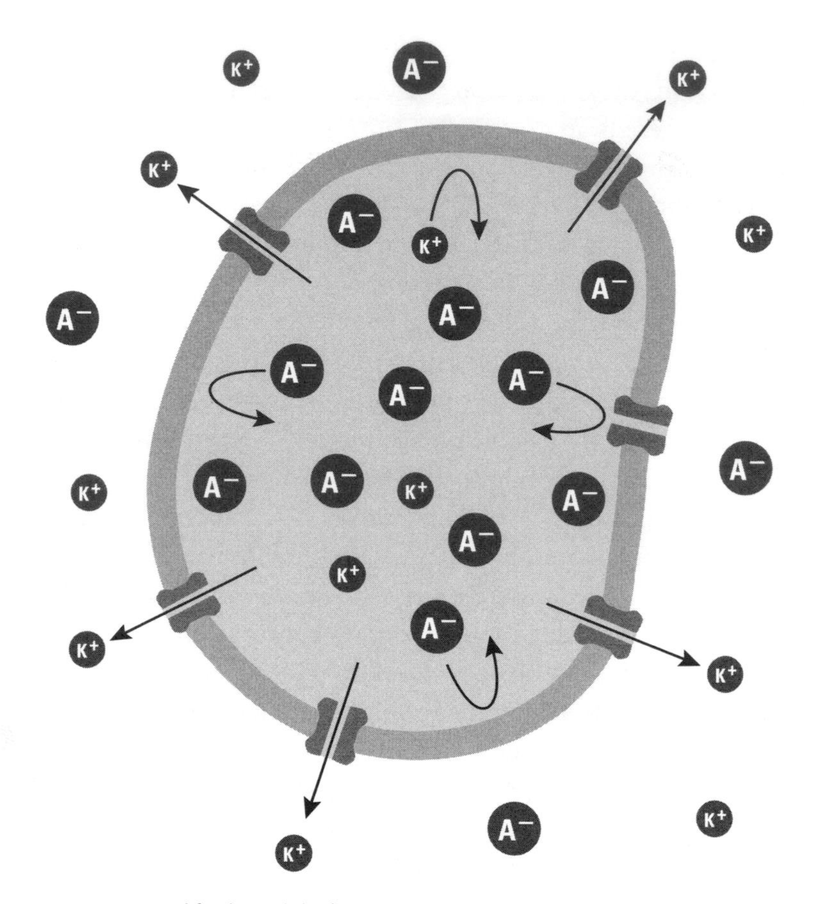

Figure 9 A simplified model of a cell showing how a resting potential is established. The concentrations of K^+ and A^- are higher inside the cell than outside, thus diffusion pressure to move from inside to outside the cell exists for both ions. Channels are present that let K^+ cross the membrane but none that allow A^- to pass. Some K^+ ions exit the cell, causing the buildup of negative charge inside and positive charge outside. This difference in charge is the resting potential.

Establishment of the resting potential across a cell membrane is somewhat more involved that the simple situation just described, but K^+ is the principal ion that determines the resting potential of virtually all nerve cells, and the principles illustrated by this example are correct. If K^+ levels on the outside of nerve cells rise and thereby reduce the difference in K^+ concentration between inside and out, the pressure on K^+ ions to move from inside to outside drops. The result is a smaller resting voltage across cell membranes, resulting in the generation of abnormal electrical signals by nerve and muscle cells. This happens in kidney disease, as was described at the beginning of the chapter: K^+ levels in the blood rise, which means that nerve and muscle cells fail to maintain normal resting potentials and to generate proper electrical signals. The heart is especially affected, and unless the K^+ levels in the blood are lowered—by treatment with an artificial kidney (a process called dialysis), for example—death will ensue because the heart cells no longer produce coordinated electrical signals.

Neuronal Signaling

Two types of electrical signals are generated by neurons: *receptor* and *synaptic potentials* and *action potentials*. Receptor and synaptic potentials are similar and are evoked when a sensory stimulus or a synaptic input impinges on a neuron. These potentials in turn generate action potentials, which carry information along axons. *Figure 10* presents a simple scheme illustrating these relations. Sensory stimuli produce receptor potentials in cells, and the receptor potentials generate action potentials along an axon. When an action potential reaches a synapse, a chemical is released and produces a synaptic potential in the adjacent neuron. This in turn generates an action potential in the axon of that neuron, and the information travels on to yet another cell. Eventually, neurons in the chain will impinge on effector cells, such as muscle cells, and a behavior will result.

Receptor and synaptic potentials have many common properties and can be considered together. Our model will be a synaptic potential that enhances activity in a contacted cell. Synapses producing such potentials are called *excitatory synapses*. (Synaptic potentials that depress the activity of a cell also occur widely in the brain; they are produced at *inhibitory synapses*.) At excitatory synapses, specific channels are found in the membrane (termed the *postsynaptic membrane*) across from where the chemical is released. In the absence of released chemical, the channels in the postsynaptic membrane are closed. But when the synapse is active, the chemical floods the space between the two cells, interacts with the channels, and causes them

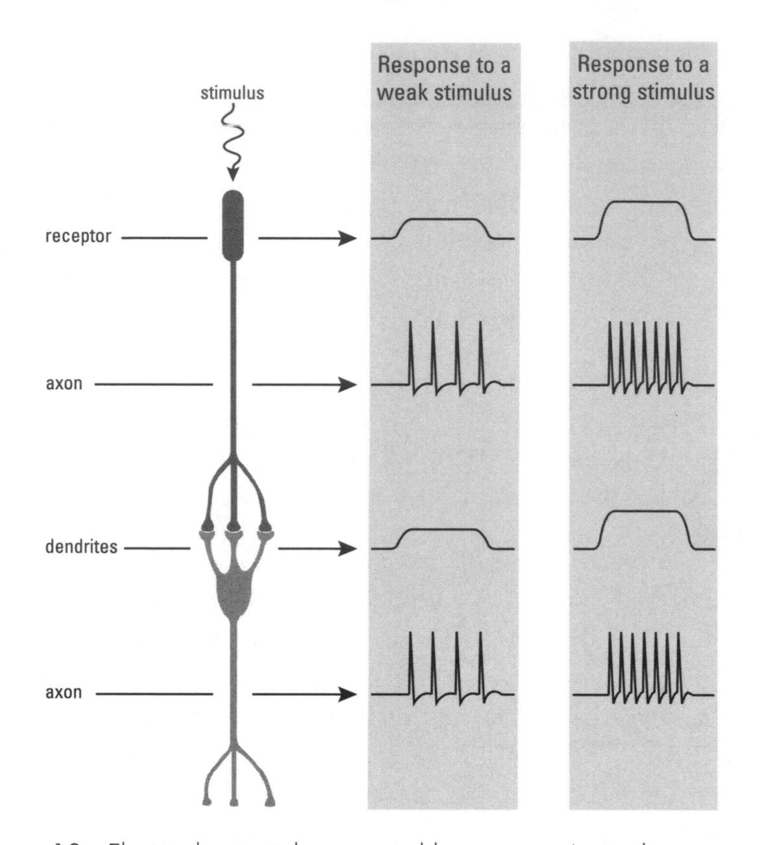

Figure 10 Electrical potentials generated by neurons. A stimulus to a receptor (left) elicits receptor potentials (right) whose amplitudes are graded (larger or smaller) depending on stimulus strength. Transient action potentials, whose amplitude is constant but whose frequency alters in response to stimulus strength, are generated along the receptor's axon. Synapses between the receptor and adjacent neuron (left) results in the generation of graded synaptic potentials in the dendrites of the neuron (right). Information is transmitted down the neuron's axon via action potentials.

to open and allow mainly Na^+ to pass across the membrane. Since Na^+ has a positive charge, the inside of the postsynaptic cell becomes more positive as Na^+ flows in. Similar excitatory signals to a cell are seen in response to both sensory and synaptic input to cells.

The stronger the input, the larger the signal generated; that is, more ions flow into the cell and the voltage change across the membrane is greater. Thus, the strength of the input signal is coded in terms of the amount of voltage change; this is a graded signal. Changes in membrane voltage of 10–50 millivolts can occur at excitatory synapses and at sensory receptor sites.

The second type of electrical signal generated by neurons, the action potential, also results from Na^+ flowing through channels in the membrane. Yet these channels open in response not to transmitter or a sensory stimulus but to a change in voltage across the membrane. At resting membrane voltage (when a neuron's membrane potential is 70 millivolts inside negative, or –70 mV), the channels are closed, but when the membrane voltage becomes more positive, the channels begin to open and admit Na^+.

As the membrane voltage becomes more positive, more Na^+ channels open, letting more Na^+ into the cell. As Na^+ comes across the membrane, the inside of the cell becomes more positive still, which opens even more Na^+ channels, and soon all the Na^+ channels in that part of the membrane rapidly open. This is a positive-feedback (or regenerative) system that results in a substantial change in membrane voltage (about 100 millivolts) occurring in less than a millisecond (1/1000 of a second). The voltage-sensitive channels, once opened, do not stay open very long. Within another millisecond or so, they close spontaneously, and recovery of membrane voltage to resting levels then occurs over the next 2 milliseconds or so.

This second type of potential, termed the action potential, is typically generated in a neuron close to where its axon arises, and is an all-or-nothing electrical event. Once initiated, the voltage change across the membrane is always the same size. To initiate an action potential requires a change of membrane voltage of about 15 mV, but once this threshold level is reached, every action potential generated is exactly the same size. In neurons, the membrane voltage change that generates action potentials is provided by the synaptic and receptor potentials; hence the term *generator potentials*.

If all action potentials are exactly the same size, how do they code for stimulus strength? Not by size of event but by frequency. The stronger the stimulus to a neuron, the more action potentials are generated per unit time. A weak stimulus generates few action potentials, a strong one many more action potentials per unit time. *Figure 11* shows

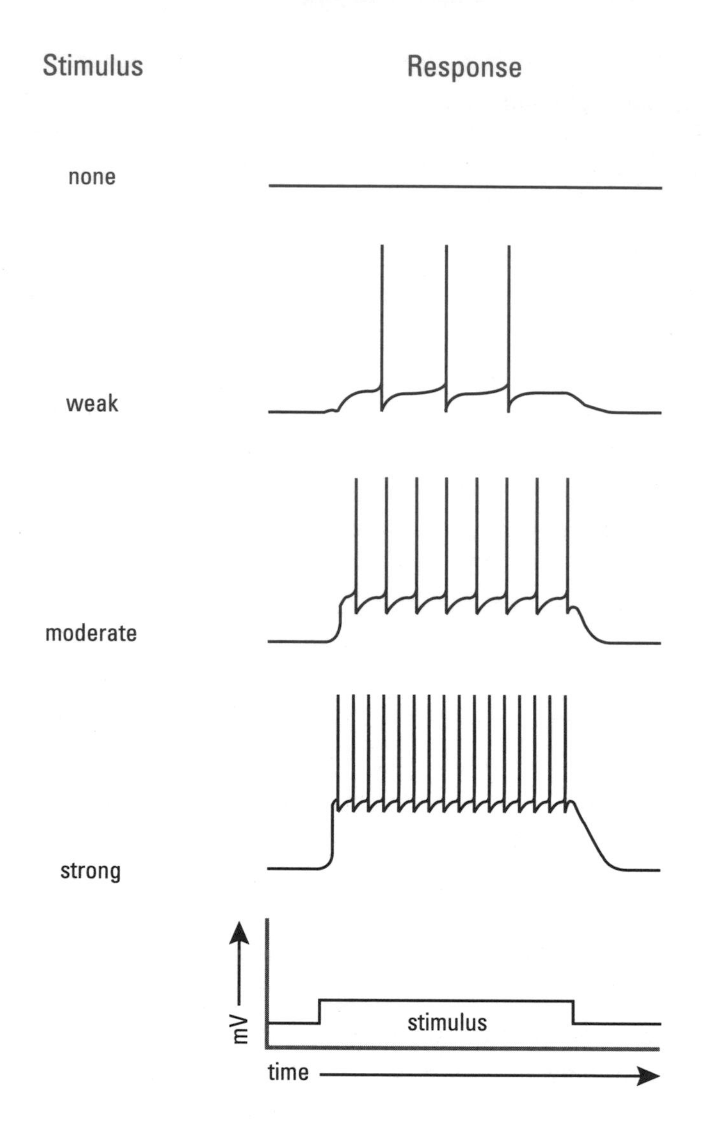

Figure 11 The relation between receptor-synaptic potentials and action potentials. The greater the amplitude of the receptor-synaptic potential, the greater is the frequency of the action potentials generated. Weak stimuli generate small receptor-synaptic potentials and correspondingly fewer action potentials are generated. Strong stimuli generate large receptor-synaptic potentials and many more action potentials.

the relations between receptor-synaptic and action potentials in the nervous system. In receptors and in dendrites, potentials are generated whose sizes are graded according to stimulus strength; along axons, action potential frequency codes stimulus strength. In receptors and in dendrites, therefore, information is coded by an amplitude modulation (AM) system whereas axons code information by a frequency modulation (FM) system. Radio signals are also coded in these two ways—certain radio stations send AM signals, others FM signals—and with the flick of a switch, most radios can receive either type of signal.

Transmission down Axons

Why do axons generate action potentials—or why do neurons generate two types of potentials? The answer is simple. Action potentials propagate themselves, but receptor and synaptic potentials do not. This means that once an action potential is generated, it moves along the axon without a change in amplitude. It remains the same size from the beginning to the end of the axon, even if the axon is several feet long! The axon potential is thus ideal to carry information long distances, which in the nervous system is a distance of more than 1–2 millimeters. Receptor and synaptic potentials, though, become smaller and smaller as they move away from their generation site.

How does the action potential propagate itself along an axon? This results simply from the way action potentials are generated; that is, an action potential is initiated when a change of about 15 mV occurs across a bit of the cell membrane. But once the action potential is initiated, it rapidly becomes full-size, because more and more Na^+ channels open. The Na^+ passing through the channels quickly makes the cell more positive inside. One consequence of this is to make the voltage along the next bit of membrane more positive. As this happens, Na^+ channels in that part of the membrane begin to open, resulting in further voltage change, and soon threshold is reached (*Figure 12*). Hence, action potentials are generated all along the length of an axon once a spike is initiated in the axon.

An interesting question is why spikes travel only in one direction along axons. The reason is that once a bit of membrane has generated an axon potential, that bit of membrane cannot be excited again for a few milliseconds. This is a long enough time so that the action potential moving down the axon is sufficiently far away not to affect that part of the membrane previously active.

Action potentials travel down axons at rates of 100–200 miles an hour in animals such as ourselves. In invertebrates (animals without backbones), action potentials travel at much slower speeds, 30–40 miles an hour. Glial cells contribute to the rapid rate of action potential travel down vertebrate axons by forming an insulating layer, called

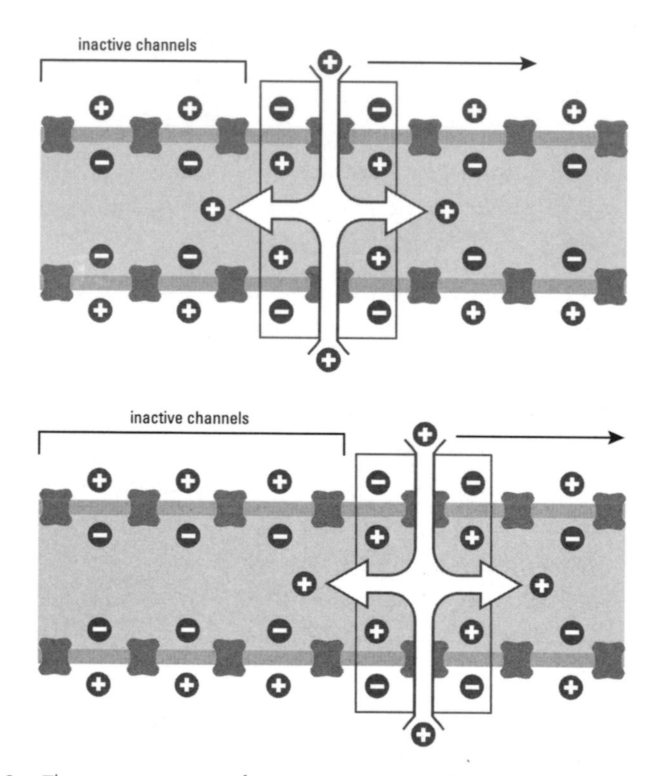

Figure 12 The propagation of action potentials along an axon. When an action potential is generated (boxed area), positively charged Na^+ ions flow into the axon. This results in positive charges moving down the axon, making the inside of the axon more positive. This results in the opening of Na^+ channels in adjacent parts of the membrane and ultimately in the generation of another action potential. The action potential moves only in one direction (to the right) because channels recently activated (to the left) are inactive for a short time—they will not open in response to changes in membrane voltage.

myelin, around axons (*Figure 13*). They do this by wrapping many layers of membrane around the axons. The myelin membrane has little protein in it, hence few channels. Thus ions cannot easily cross myelin; it acts as a good insulator. Myelin not only speeds action potential conduction, but it makes action potential propagation more efficient by limiting new action potentials to certain regions—nodes—along the axon where there are gaps in the myelin. Action potentials are not generated continuously along vertebrate axons; rather they are generated only at the nodes. So action potential generation jumps from node to

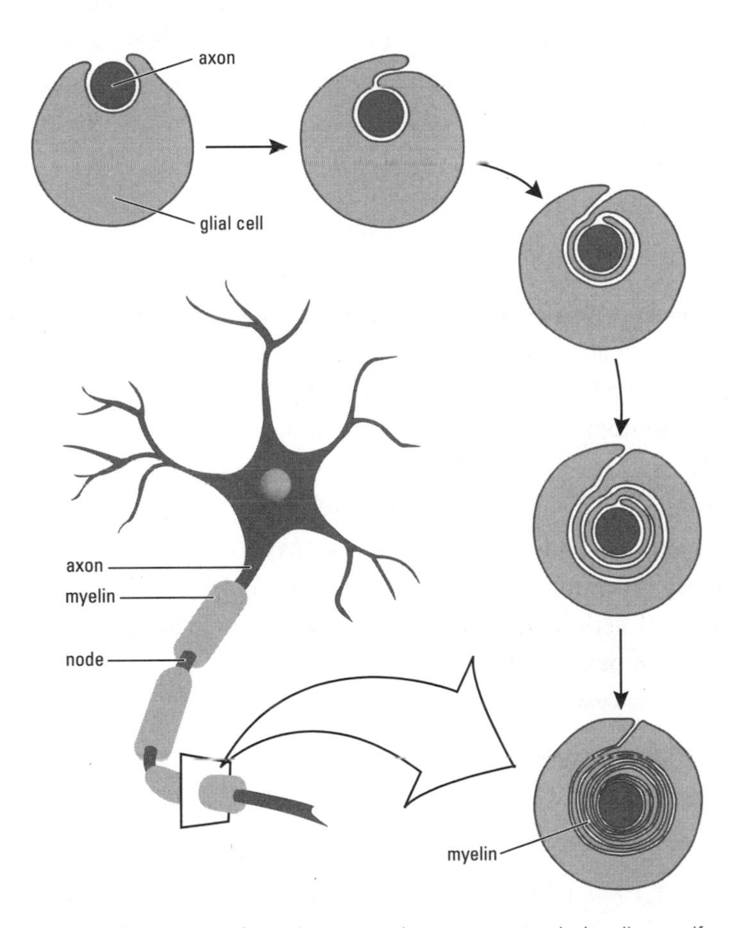

Figure 13 The formation of myelin around an axon. A glial cell engulfs an axon and then wraps many layers of membrane around the axon. The cytoplasm between the layers of membrane gets squeezed out, leaving a highly compacted myelin sheath. The myelin sheath is not continuous along an axon but is interrupted by gaps or nodes, as shown in the drawing at bottom left.

node, and this works so long as the nodes are not too far apart. The advantage is that much less energy is needed to transmit action potentials down vertebrate axons as compared with invertebrate axons (which do not have myelin), and action potentials travel much faster down vertebrate axons.

If myelin is damaged or diseased, propagation down axons is compromised, which happens in multiple sclerosis (MS) and several other demyelinating conditions. Multiple sclerosis is one of the most common central nervous system diseases of young adults, affecting about one in a thousand individuals. It strikes mainly between the ages of twenty-five and forty, and about two-thirds of the cases occur in women. Early symptoms include blurred vision, uncoordinated walking, numbness, and fatigue. In later stages, patients may exhibit slurred speech, tremors, memory loss, and severe paralysis.

The cause of MS is not known, but it is thought to be an autoimmune disease. For unknown reasons, the immune system of those affected forms antibodies against myelin, causing it to disintegrate. The disease is also capricious—patients typically have remissions and relapses that can occur over a period of years.

Why does MS cause neurological symptoms? When myelin is lost, action potential conduction is first slowed and eventually may be entirely blocked. In essence, the loss of myelin has the same effect as increasing the distance between the nodes would, compromising action potential generation. If enough myelin is lost, action potentials no longer can jump the gap between nodes. A slowing or even a partial block of action potential transmission down axons has a devastating effect on neural activity and behavior.

Synapses

I have so far dwelled on excitatory synapses. At such synapses, channels allow mainly Na^+ to flow into the cell, which causes the inside of the cell to become more positive and the cell membrane to approach action potential threshold. For this reason these synapses are called excitatory; they cause the cell to generate action potentials.

But, as noted earlier, there are inhibitory synapses as well. These synapses make a neuron less likely to fire an action potential. How do they work? In a fashion similar to excitatory synapses, but at inhibitory synapses, a different ion flows across the membrane. Channels at inhibitory synapses typically allow Cl^- to pass into the cell; since Cl^- is negatively charged, the cell becomes more negative inside when inhibitory synapses are activated. This causes membrane voltage to move away from the action potential threshold. Inhibitory synapses thus slow down or prevent action potential generation.

The rate of action potential generation by a neuron depends on the interplay of excitatory and inhibitory synapses onto the cell, as shown schematically in *Figure 14*. At the top of the figure, excitatory synapses impinge on the cell's dendrites; inhibitory synapses are on the cell body. The action potentials are first generated by the membrane adjacent to the axon. Surrounding the axon is myelin, interrupted by nodes that generate the action potentials traveling down the axon.

In the bottom part of *Figure 14* is a record of membrane voltage, as would be recorded by placing an intracellular recording probe into the cell and making measurements over time. When the probe enters the cell a resting potential of about –70 mV is recorded. When excitatory input to the neuron is applied (depicted as ticks along the line marked "excitatory"), small synaptic potentials are elicited in the dendrites. These add together, and with repetitive excitatory stimuli they cause a buildup of positive charge within the cell that brings membrane potential to threshold for action potential generation.

So long as excitatory input continues, the cell will generate action potentials—one after another. But if inhibitory input to the cell is activated (ticks along the line marked "inhibitory"), the membrane potential becomes more negative. The membrane potential falls below action potential firing threshold levels and the cell stops firing. Once the inhibitory input stops, the excitatory input drives the membrane potential to the action potential threshold again, and the cell generates action potentials as long as excitatory input continues. Thus, the rate of spike firing in a neuron reflects the balance of excitatory and inhibitory input to a neuron, and the message traveling down the axon

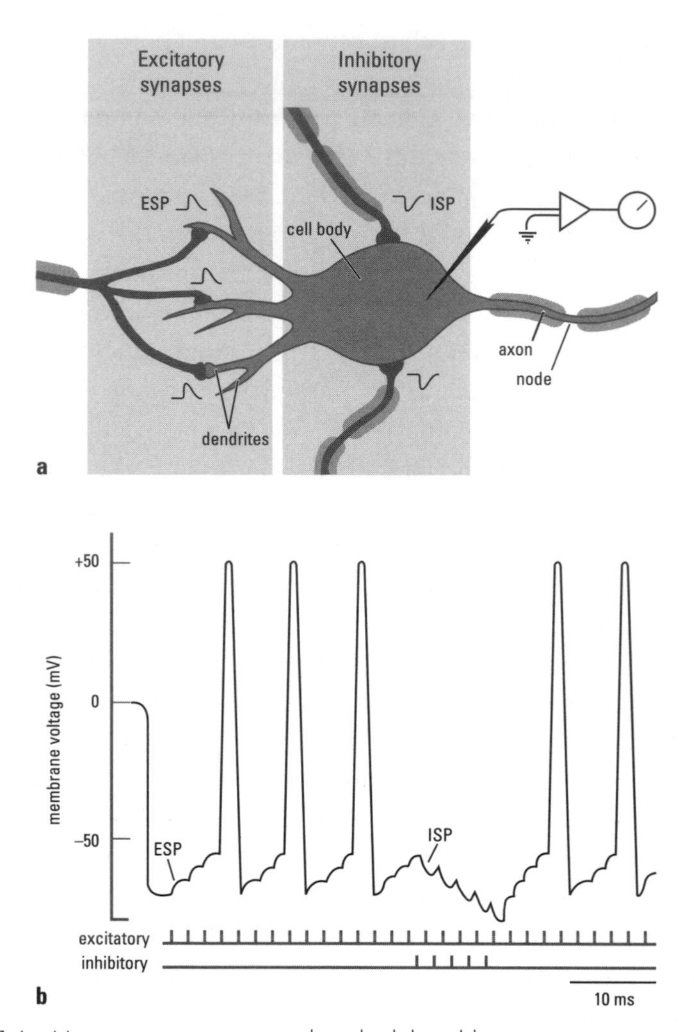

Figure 14 How a neuron is excited and inhibited by synaptic input. *a*: Excitatory synapses are found predominantly on dendrites, inhibitory synapses on the cell body. Each active synapse generates a small positive excitatory synaptic potential (ESP) in the dendritic tree or a small negative inhibitory synaptic potential (ISP) in the cell body. *b*: A recording made from the cell body will record the resting, synaptic, and action potentials. The resting potential of the cell is approximately −70 mV. Excitatory input elicits excitatory synaptic potentials (ESPs) which sum, causing the membrane potential to become more positive. This results in the generation of action potentials. Activating the inhibitory input elicits inhibitory synaptic potentials (ISPs), which cause the membrane potential to become more negative. As this happens the neuron ceases firing. Once the inhibitory input stops, the cell membrane becomes more positive and again fires action potentials.

reflects this balance.

Synaptic Mechanisms

Among the key structures at a synapse are the *synaptic vesicles* that store the chemicals to be released, which are called *neurotransmitters.* When a synapse becomes active, the vesicles join to the membrane and release the stored neurotransmitter. How does this happen?

When an action potential travels down an axon and reaches a synaptic terminal, the membrane surrounding the terminal becomes more positive because of the Na^+ flowing across the membrane through Na^+ channels. In the terminal membrane are other channels that respond to this voltage change; they open and admit an ion that has two extra positive charges, calcium (Ca^{2+}). Calcium ions, in ways still not well understood, promote the docking of vesicles to the membrane (*Figure 15*). This results in the fusion of the vesicle to the membrane and the opening of the vesicle to the outside. Neurotransmitter is then released and flows to channels on the postsynaptic membrane, thus activating them.

Figure 15 shows two other important features of synaptic transmission. First, after the synaptic vesicles bind to the presynaptic membrane and release their contents, they become incorporated into the terminal membrane. But away from the active zone of the terminal (where neurotransmitter is released), new vesicles are formed by the inpocketing of membrane. Thus, vesicles recycle in the synaptic terminal.

The other important aspect of synaptic transmission illustrated in the figure is the breakdown or reuptake of neurotransmitter. Once neurotransmitter is released from the vesicles, it is necessary to get rid of it rapidly so new neurotransmitter can have effects. This occurs in two ways. Some neurotransmitters are broken down by enzymes to inactive products; most neurotransmitters, though, are rapidly taken back into the terminal by membrane proteins called *transporters*. These transporters move the neurotransmitter into the terminal, where it is repackaged into the newly formed vesicles. There is much interest in transporters among neuroscientists because potent drugs such as cocaine and Prozac inhibit transporter function. Inhibition of a transporter allows a neurotransmitter to remain in the synaptic cleft for a longer period of time. In essence, it is as

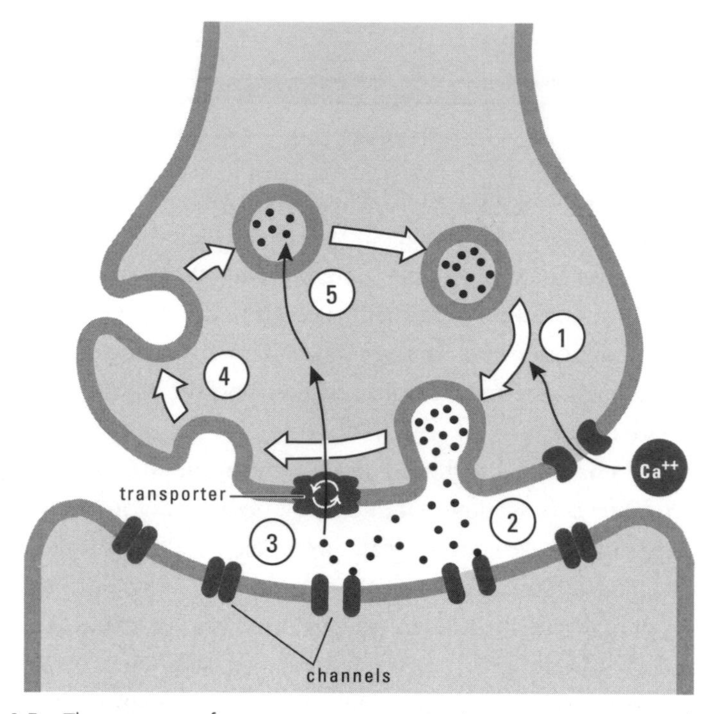

Figure 15 The process of synaptic transmission. A synapse is activated when the terminal membrane becomes more positive as a result of an action potential reaching the terminal. *1*: This causes voltage-sensitive Ca^{2+} channels to open, admitting Ca^{2+} into the terminal. Ca^{2+} facilitates the binding of synaptic vesicles to the membrane. *2*: The vesicles release their neurotransmitter content, which diffuses to the postsynaptic membrane and activates channels there. *3*: Neurotransmitter is either broken down by enzymes (not shown) or taken back up into the terminal by transporters, allowing for its reuse. *4*: After fusing with the membrane, the synaptic vesicles become incorporated into the membrane. New synaptic vesicles form by an inpocketing of the terminal membrane. *5*: The newly formed synaptic vesicles are filled with neurotransmitter and the process is ready to go again.

though one has increased the amount of neurotransmitter released at the synapse, and there is now abundant evidence that altering the levels of neuroactive substances at synapses can cause specific psychological changes.

Vulnerability of Synapses

Synapses are vulnerable to blockage, and many drugs and poisons that affect the nervous system disrupt one or another aspect of their ability to transmit or receive signals. Indeed, drugs affect virtually every aspect of synaptic transmission. For example, a toxin from the deadly bacterium *Clostridium botulinum* that causes botulism potently blocks synaptic transmission by preventing the release of neurotransmitter from synaptic terminals. The toxin does this by interfering with the binding of the synaptic vesicles to the membrane of the terminal. Proteins present in both the synaptic vesicle and cell membranes are involved in the attachment and fusion of the vesicles to the cell membrane that results in the release of neurotransmitter at a synapse. The botulinum toxin acts by breaking down one of these attachment proteins. Thus, the vesicles cannot fuse to the membrane and release their neurotransmitters.

Clostridium botulinum is found commonly in soil and on fruits and vegetables. It is an anaerobic bacterium—that is, it thrives and multiplies under conditions of low oxygen. It can become a serious problem to humans who eat canned fruits and vegetables that have not been heated enough during canning to destroy the bacterium. If any bacteria remain in the low-oxygen atmosphere of a can or jar, they multiply and produce copious amounts of the toxin. Eating the tainted food can cause death by blocking synaptic transmission throughout the nervous system; a person so poisoned cannot breathe, for example. Botulinum toxin poisoning was much more of a problem in the early days of canning, but even today occasional incidents occur. Botulinum toxin can be made fairly easily and is exceptionally potent; the worry now is that it may be used as a biological warfare agent. One estimate predicts that as many as 40,000 people could be killed by just 200 pounds of botulinum toxin. Tetanus toxin is related to the botulinum toxin and also acts by preventing the release of neurotransmitters at synapses.

Synaptic transmission can also be blocked by preventing a neurotransmitter from interacting with the channels in the postsynaptic membrane. Curare is one example of substances that have this effect. It is a paralyzing agent that blocks synaptic transmission between nerve and muscle at the neuromuscular junction, rapidly causing paralysis and death. This naturally occurring substance, found in certain climbing vines, was discovered by South American Indians, who used it to coat the tips of arrows and spears. Another agent that blocks the neuromuscular junction by preventing neurotransmitter from interacting with membrane channels is α-bungarotoxin, a deadly component of cobra snake venom. It binds tightly to the channels, essentially destroying them. Once α-bungarotoxin attaches to a channel, transmitter can no longer interact with the channel and activate it. Like curare, α-bungarotoxin causes muscle paralysis and death.

Agents also can prevent the breakdown of neurotransmitter after its release or prevent reuptake of the transmitter into the terminal. Examples of the former are certain organic phosphates, which are the principal ingredients in many pesticides and in the deadly nerve gases such as sarin, the gas that killed a dozen people in the Tokyo subway station terrorist attack in 1995. A tiny amount of these agents causes very rapid paralysis by blocking the enzyme that breaks down the neurotransmitter at the neuromuscular junction. Neurotransmitter builds up excessively at the synapse and the synapse can no longer function properly. Another drug that acts the same way is eserine, a derivative of the African calabar bean. Eserine was used in some parts of Africa as a truth serum. The legend is that a person who was unjustly accused would rapidly drink down the mixture. Since eserine also causes vomiting, such an individual would vomit much of the poison and survive. The guilty individual would drink the truth serum more slowly, would absorb more of the poison, and would die.

Some substances prevent or inhibit the reuptake of neurotransmitters. As noted above, neuroactive drugs such as cocaine and Prozac exert their effects by inhibiting the transporters involved in reuptake of neurotransmitters. If reuptake is entirely prevented, synaptic transmission will soon be completely shut down as neurotransmitter floods the synaptic cleft, and dire consequences will result. Neuroactive drugs such as cocaine and Prozac inhibit reuptake processes only partially—they simply raise neurotransmitter levels in the synaptic cleft. Furthermore, they affect only the uptake of certain substances released at synapses.

A Synaptic Disease: Myasthenia Gravis

A disease of the synapses between nerve and muscle, myasthenia gravis, is characterized in its early stages by muscle weakness and fatigue. Diplopia (double vision), due to a weakness of the ocular muscles, in another early symptom, along with droopy eyelids. In its latter stage, patients may be bedridden and even die. Fortunately, myasthenia gravis is a relatively rare disease; it affects about 25,000 people in the United States. The diagnosis, treatment, and analysis of this disease have depended on a detailed understanding of what happens during synaptic transmission at the neuromuscular junction and the drugs that block transmission at this synapse.

Many years ago, a patient suffering from myasthenia gravis was given a tiny amount of curare. The drug induced muscle weakness and fatigue far beyond that expected based on observations of normal people given similar amounts of the drug. These findings provided some of the first evidence that myasthenia gravis is a disease involving the synapses between nerve and muscle. The excessive sensitivity of myasthenia gravis patients to drugs such as curare has been used to diagnose the disease. Yet giving a patient eserine, which prevents breakdown of the neurotransmitter at the neuromuscular junction, is used to help patients with the disease. Their symptoms are alleviated, and, indeed, eserine and eserine-like drugs are the primary therapy for the disease.

Why would curare accentuate and eserine relieve symptoms of the disease? Experiments in the mid-1960s showed that patients with myasthenia gravis have fewer channels in the postsynaptic membrane at muscle synapses than do normal individuals. Curare, by interfering with the interaction of neurotransmitter with channels, exaggerates the deficit by reducing the number of activated channels. Eserine, by contrast, interferes with the breakdown of neurotransmitter and allows it to remain intact longer at the synapse. Thus, in the presence of eserine, channels can be activated longer than normal. This excites the muscle more, as though more channels were in the postsynaptic membrane.

Finally, why are there fewer neurotransmitter-activated channels in patients with myasthenia gravis? There is now good evidence that myasthenia gravis, like multiple sclerosis, is an autoimmune disease. Individuals with myasthenia gravis form antibodies against the channels in the postsynaptic membrane. Why this happens is not understood, but what is known is that the antibodies attach tightly to the channels leading to their destruction, much as α-bungarotoxin does. Ordinarily, neurotransmitter-activated channels are synthesized continuously by muscle (and nerve) cells, replacing old ones or ones that have been damaged for one reason or another. Indeed, neurotransmitter-activated channels at the neuromuscular junction are completely replaced in us every week or so. In myasthenia gravis, the antibodies inactivate the channels as fast as or faster than they can be made, hence the lower number of functional channels at the neuromuscular junction in these patients.

An understanding of the causes of myasthenia gravis suggests a strategy for its cure. If antibody levels in the blood can be reduced, the number of functional channels in the postsynaptic membrane should increase. It is possible to reduce antibodies in a patient's blood by treatment with the artificial kidney (dialysis), and spectacular improvement is seen in such patients. Unfortunately, the effect is only temporary; within a few days after dialysis, antibody levels increase. At present there is no way to lower antibody levels permanently, but if this active area of research can be made successful, patients with myasthenia gravis and other autoimmune diseases could be completely cured.

CHAPTER 3

Drugs and the Brain

Tess was the eldest of ten children born to a passive mother and an alcoholic father in the poorest public-housing project in our city. She was abused in childhood in the concrete physical and sexual senses which everyone understands as abuse. When Tess was twelve, her father died, and her mother entered a clinical depression from which she never recovered. Tess—one of those inexplicably resilient children who flourish without any apparent source of sustenance—took over the family. She managed to remain in school herself and in time to steer all nine siblings into stable jobs and marriages.

Her own marriage was less successful. At seventeen, she married an older man, in part to provide a base outside the projects for her younger brothers and sisters, whom she immediately took in. She never went to the movies alone with her husband; the children came along. The weight of the family was always on her shoulders. The husband was alcoholic, and abusive when drunk. Tess struggled to help him stop drinking, but to no avail. The marriage soon became loveless. It collapsed once the children—Tess's siblings—were grown and one of the marriage's central purposes had disappeared.

Meanwhile, Tess had made a business career out of her skills at driving, inspiring, and nurturing others. She achieved a reputation as an administrator capable of turning around struggling companies by addressing issues of organization and employee morale, and she rose to a high level in a large corporation. . . .

That her personal life was unhappy should not have been surprising. Tess stumbled from one prolonged affair with an abusive married man to another. As these degrading relationships ended, she would suffer severe demoralization. The current episode had lasted months, and, despite a psychotherapy in which Tess willingly faced the difficult aspects of her life, she was now becoming progressively less energetic and more unhappy. It was this condition I hoped to treat, in order to spare Tess the chronic and unremitting depression that had taken hold in her mother when she was Tess's age.

Though I had learned some of this story before my consultation with Tess, the woman, when I met her, surprised me. She was utterly charming. . . . She was a pleasure to be with, even depressed. I ran down the list of signs and symptoms, and she had them all: tears and sadness, absence of hope, inability to experience pleasure, feelings of worthlessness, loss of sleep and appetite, guilty ruminations, poor memory and concentration. . . .

Two weeks after starting Prozac, Tess appeared at the office to say she was no longer feeling weary. In retrospect, she said, she had been depleted of energy for as long as she could remember, had almost not known what it was to feel rested and hopeful. She has been depressed, it now seemed to her, her whole life. She was astonished at the sensation of being free of depression.

She looked different, at once more relaxed and energetic—more available—than I had seen her, as if the person hinted at in her eyes had taken over. She laughed more frequently, and the quality of her laughter was different, no longer measured but lively, even teasing.

With this new demeanor came a new social life, one that did not unfold slowly, as a result of a struggle to integrate disparate parts of the self, but seemed, rather, to appear instantly and full blown.

—Excerpted from Peter D. Kramer, *Listening to Prozac*

At synapses, neurons can not only be excited or inhibited, they can also be modified or modulated in a variety of ways. This second type of synaptic action is called neuromodulation. *I have already described how neurons are excited or inhibited at synapses; now I'll detail how neurons are modulated by synaptic action. Why are these neuromodulatory processes important? Because long-term changes that occur in the brain and result in phenomena such as remembering and learning are thought to happen because of neuromodulatory synaptic action. But also, many chemicals that alter mental states, such as LSD and the amphetamines, and diseases that cause thought disorders, such as schizophrenia, can be related to alterations in neuromodulatory synaptic transmission. And as the story of Tess described above illustrates, inducing alterations in levels of neuromodulatory substances in the brain by drugs such as Prozac can provide striking relief from serious psychiatric disorders such as deep depression.*

We know that transmission at most brain synapses is chemical. A substance released from one neuron diffuses to an adjacent neuron and induces changes in that neuron. A few synapses are electrical; at such synapses the cell bodies of two adjacent neurons become closely apposed and protein channels create a bridge between the contacting cells. These channels enable ions to flow directly from one cell to another. A change in membrane voltage in one cell results in a rapid change in voltage in the contacted cell as the ions flow from one cell to the next. Such transmission is very fast and can go in either direction. Electrical synapses, then, synchronize activity between neurons or permit reciprocal interactions between neurons.

But signals cannot be amplified at electrical synapses—amplification is possible only at chemical synapses. Also, signal polarity cannot be changed at electrical synapses—from excitation in one cell, for example, to inhibition in the adjacent cell or vice versa. Not only that, there is little in the way of signal modification at electrical synapses. Thus, chemical synapses are predominant in the brain because they allow for a richness of interactions between neurons.

Neurotransmitters and Neuromodulators

As many as fifty different substances may be released at chemical synapses in the brain. Usually no more than one or two substances are released at any one synapse, and it is generally believed that all the synapses made by a neuron release the same substance or substances. So neurons have a lot of diversity and specificity in relation to the substances they release at their synapses.

It is convenient to classify the substances released at synaptic sites into two categories, neurotransmitters and neuromodulators, based on their mode of action on postsynaptic cells. *Neurotransmitters* are substances that interact directly with channels in the postsynaptic membrane; they mediate fast excitatory or inhibitory synaptic action. When a neurotransmitter activates such channels, they open and allow certain ions to cross the membrane. As you already know, when positively charged ions such as Na^{+} cross the membrane, the inside of the cell becomes more positive and the neuron is more likely to generate an action potential—it is excited. Conversely, if negatively charged ions such as Cl^- cross the membrane, the inside of the cell becomes more negative and the cell is less likely to generate an action potential—it is inhibited. These excitatory and inhibitory interactions are quite fast. Once a synapse is activated, it takes less than half a millisecond (0.0005 second) for a voltage change to occur in the postsynaptic cell. Furthermore, the change in voltage in the postsynaptic cell lasts only a short time—between 10 and 100 milliseconds, or no more than 0.1 second.

Neuromodulators act quite differently on neurons. When neuromodulators are released at synapses, they diffuse to the postsynaptic membrane where they interact with membrane proteins called *receptors*, which are linked to intracellular enzyme systems. Activating these receptors does not directly open channels in the membrane; rather, the neuromodulators' action is mediated by biochemical changes in the postsynaptic neuron. In a typical situation, intracellular enzymes are activated and synthesize small second-messenger molecules (the first messenger is the neuromodulator). Many aspects of neural cell function can then be altered, from the properties of the channel proteins in the membrane to the expression of genes in the nucleus. In this manner, profound physiological and structural changes can happen in a neuron as a result of neuromodulation.

Neuromodulatory effects typically have a slow onset, usually seconds, and then can last for long periods of time—minutes, hours, days, or even longer. As noted above, the evidence is accumulating that long-term changes in the brain, which underlie phenomena such as memory and learning, result from neuromodulatory synaptic interactions.

Some substances released at synapses act exclusively as neurotransmitters; others are solely neuromodulators. Most substances released at synapses can act as both. At some synapses, they interact with membrane channels postsynaptically, while at other synapses they interact with membrane receptors linked to intracellular enzyme systems. *Acetylcholine* is a typical example. It is the neurotransmitter used at all synapses between nerve and muscle in vertebrates (the neuromuscular junction), but it can also activate membrane receptors in the brain linked to biochemical systems. These two kinds of actions can be distinguished by pharmacological means; certain drugs will block one action but not the other, and vice versa. Curare blocks the action of acetylcholine at the neuromuscular junction, whereas the drug atropine blocks the effect of acetylcholine on membrane receptors. Atropine has no effect on the neuromuscular junction, and curare does not block the neuromodulatory action of acetylcholine on membrane receptors. Some agents can specifically activate one or the other action of acetylcholine. Nicotine activates the membrane channels specific to acetylcholine, while muscarine activates acetylcholine receptors linked to enzyme systems. These two effects are often referred to as the nicotinic or muscarinic actions of acetylcholine.

Figure 16 shows how a neuromodulatory system works. The system chosen for illustration uses an enzyme (adenylate cyclase) that synthesizes a second-messenger molecule called cyclic AMP. A neuromodulator (the first messenger), released from vesicles in the presynaptic terminal, interacts with receptors on the postsynaptic membrane and thereby activates them. The activated receptors now interact with an intermediate protein, called a *G-protein*, which in turn interacts with adenylate cyclase and activates it.

The Nobel Prize for Physiology and Medicine was given to Alfred Gilman and Martin Rodbell in 1994 for the discovery of G-proteins. Virtually all known membrane receptors linked to intracellular enzyme systems—including the light-sensitive molecules in photoreceptors cells, the odorant-sensitive proteins in olfactory cells, and the neuromodulatory receptors on neurons—interact first with G-proteins. There are a number of G-proteins known that can activate (or inhibit) a variety of intracellular enzymes.

Regarding the system in *Figure 16*, activated adenylate cyclase promotes the conversion of the molecule *adenosine triphosphate* (ATP), which is ubiquitous in cells and stores and provides energy, into a smaller molecule called *cyclic AMP*. Cyclic AMP exerts its effects by activating another enzyme, termed a *kinase*, that adds a phosphate group to cellular constituents, usually proteins. This process, called *phosphorylation*, is a favorite way for cells to activate or inactivate biochemical reactions or to modify proteins' properties.

The kinase activated by cyclic AMP is called *protein kinase A* (PKA).

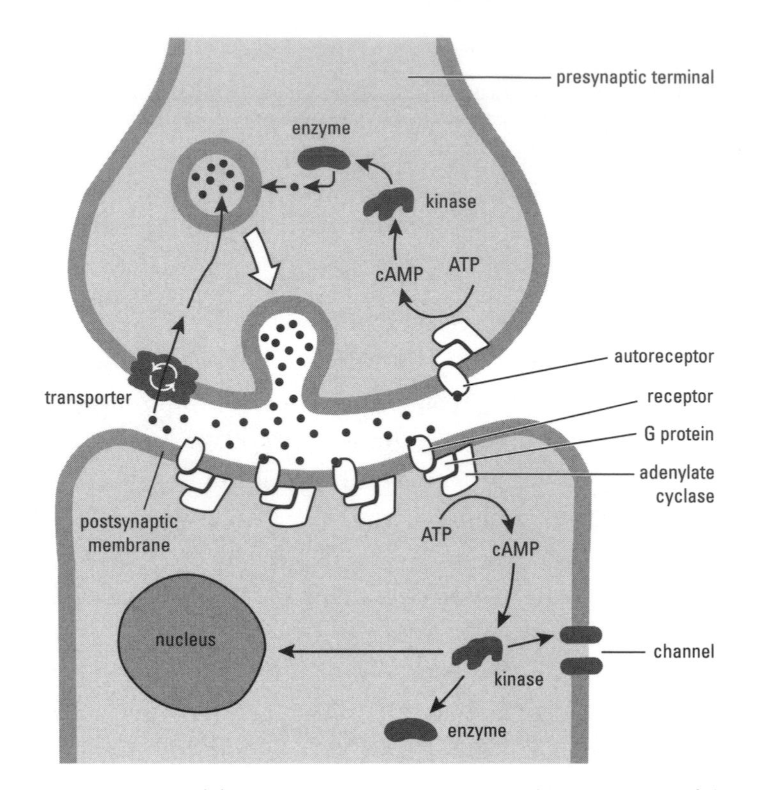

Figure 16 Neuromodulatory synaptic transmission. The activation of the presynaptic terminal, binding of vesicles to the terminal membrane, release of neuroactive substances from the synaptic vesicles, and reuptake of the released substances back into the terminal by transporters are identical to the processes shown in the scheme of synaptic transmission in Figure 15. However, at neuromodulatory synapses, the released substances bind to protein receptors on the postsynaptic membrane linked to a G-protein. In the case illustrated, the G-protein activates an enzyme, adenylate cyclase, that converts ATP to cyclic AMP. Cyclic AMP in turn activates a kinase that can act on channels in the membrane, on enzymes in the cytoplasm, or on proteins regulating gene expression in the nucleus. In addition, autoreceptors, found on the presynaptic terminal, may be activated. In the case shown, the cyclic AMP cascade leads to the activation of an enzyme involved in synthesizing the neuroactive substance released from the synaptic vesicles of the terminal.

It can cause effects at many levels of the cell, including the nucleus, the cytoplasm, and the membrane. In the nucleus, genes can be turned on or off; in the cytoplasm, enzymes can be activated or inactivated, including those involved in protein synthesis; and at the membrane, ion channels and other membrane proteins can be altered. For example, phosphorylation of a channel can modify its sensitivity to a neurotransmitter, or how long the channel remains open after its activation by a neurotransmitter, or even its ionic specificity—that is, which ions it will admit into the cell.

Another interesting feature of neuromodulation is that a substance released at a synapse can affect the terminal releasing it; thus, synaptic terminals can have their own receptors, termed *autoreceptors*, that respond to the neuromodulator. In our example in *Figure 16*, autoreceptors are linked to the cyclic AMP cascade. The activated kinase may, for example, phosphorylate enzymes that synthesize the substance released by the terminal. With such mechanisms a neuromodulator can regulate how much substance is made by the terminal; that is, a terminal can modify its own properties by feedback mechanisms.

Even though several second-messenger cascades are known, the list is far from complete. Cyclic AMP is the most extensively studied second messenger and the one best known. Other second messengers activate other kinases, and any kinase is very specific with regard to the molecules it can phosphorylate. One neuromodulator at a synapse can activate more than one second-messenger pathway by way of one receptor but several G-proteins; thus, a variety of processes can be altered in a cell due to a single neuromodulatory input. Any neuron probably receives many such inputs, so the possibilities for its mode of modulation are considerable.

The distinction between synaptic neurotransmission and neuromodulation is sometimes blurred. Some ion channels in the membrane activated by neurotransmitters can admit Ca^{2+} ions into neurons, and the admitted Ca^{2+} can act as a second messenger. The way this works is that the Ca^{2+} binds to a protein within the cell, termed *calmodulin*, and the Ca^{2+}-calmodulin complex interacts with specific kinases, termed *cam-kinases*. These kinases act like other second-messenger–activated kinases, phosphorylating cellular constituents and thus activating or inactivating biochemical mechanisms. Long-term changes in vertebrate neurons that relate to memory and learning appear to be mediated by such a mechanism.

Classifying Synaptic Substances

Although as many as fifty substances may be released at synapses in the human brain, they fall into one of four chemical classes. Two of these classes act mainly as neurotransmitters (acetylcholine and amino acids), and the other two act mainly as neuromodulators (monoamines and neuropeptides). I'll start with the neurotransmitters.

Acetylcholine (ACh) is the only naturally occurring substance in its class, but other chemicals mimic ACh activity; they will activate acetylcholine channels or receptors. Acetylcholine is usually an excitatory neurotransmitter, but it can exert inhibitory effects through neuromodulatory actions. Indeed, its discovery in the early 1920s by Otto Loewi, a German neurochemist, came about in relation to a slowed heart rate. Acetylcholine is released in the heart when the vagus nerve, which regulates the heart, is active; acetylcholine interacts with a receptor on heart cells linked to a G-protein. The G-protein inhibits adenylate cyclase and thus the synthesis of cyclic AMP. The fall in cyclic AMP levels diminishes the phosphorylation of certain channels in heart cell membranes, thereby slowing down the heart (*see Figure 59* in Chapter 9). What Loewi did was to stimulate the vagus nerve in an animal and collect the fluid around its heart. He demonstrated that this fluid slowed the heart of another animal whose vagus nerve was not stimulated. It was learned later that the substance released by the vagus nerve is acetylcholine.

Like acetylcholine, the amino acids released at synapses act mainly as neurotransmitters, although some can interact with membrane receptors linked to second-messenger cascades. One amino acid, *glutamate*, serves as the major excitatory neurotransmitter in the brain; two other amino acids, *glycine* and *γ-aminobutyric acid* (GABA), are the major inhibitory neurotransmitters in the brain. Other amino acids such as aspartate, cysteine, and histamine have excitatory or inhibitory actions on neurons, but not much is yet known about their roles in synaptic transmission.

Glutamate, GABA, and glycine are closely related structurally, as you can see in *Figure 17*. The molecule glutamate becomes GABA when a carbon atom and two oxygen atoms are removed. Without its tail the molecule is glycine.

Glutamate activates two types of channels. One allows Na^+ to enter a cell, thus exciting it. The other channel allows both Na^+ and Ca^{2+} to enter the cell, which excites the neuron and also activates calmodulin and cam-kinases. Thus when a neuron possesses this channel, it can be both excited and modulated at the same synapse. We have evidence that this channel plays a role in storing memories. When we experience something, not only are our neurons excited, but they must also be modified in some way to account for memory. How this happens is not well understood, but neuromodulatory pathways activated at these glutamate synapses may underlie such long-term changes in neurons.

GABA and glycine mainly activate channels that allow Cl^- to flow into neurons. This results in inhibition of the neurons. The GABA-activated Cl^- channel is of particular interest because its properties are specifically altered by three kinds of drugs: barbiturates, benzodiazepines, and alcohol. Barbiturates and benzodiazepines are used to treat anxiety, and at low concentrations alcohol can also relieve anxi-

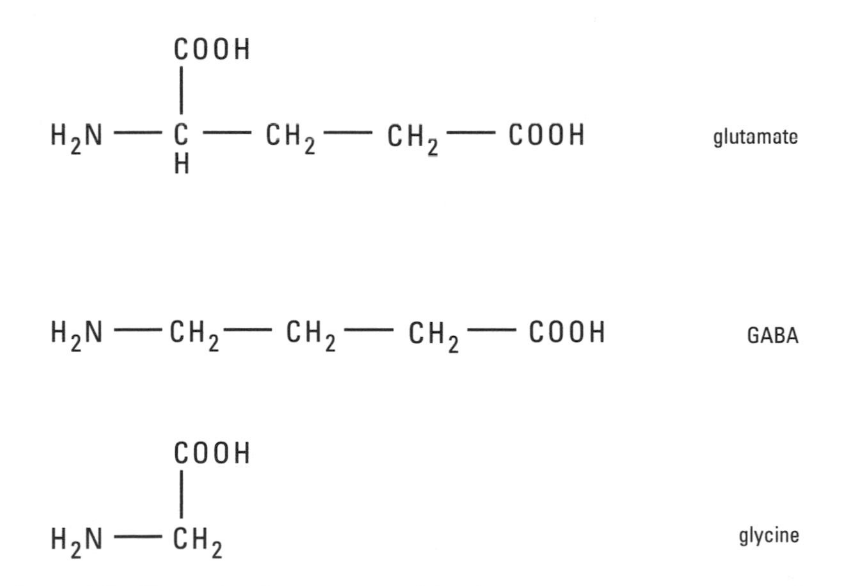

Figure 17 The molecular structures of the three most common amino acid neurotransmitters used in the brain. Glutamate, an excitatory neurotransmitter, is very similar in structure to GABA, an inhibitory neurotransmitter.

ety and promote relaxation. The benzodiazepines were introduced in the 1960s, and by the early 1980s they were used by enormous numbers of people. A variety of benzodiazepines are available; the best-known is diazepam (Valium). By the early 1980s, about 20 percent of the women and 10 percent of the men in England were taking a benzodiazepine at one time or another during the year.

Barbiturates, benzodiazepines, and alcohol all potentiate the GABA-activated Cl^- channels. When these drugs are present, the channels allow more Cl^- to cross the membrane; hence the postsynaptic neurons are more strongly inhibited. How increased inhibition relieves anxiety is not understood. What is known is that barbiturates and benzodiazepines interact at specific sites on the GABA channel, and the sites are different from the site where GABA interacts with the molecule. The fact that there exist sites specific for a drug on a membrane channel or receptor suggests that an endogenous brain substance naturally interacts with the site. As yet, no endogenous benzodiazepine or barbiturate-like substances have been identified.

Monoamines (of which there are two types—*catecholamines* and *indoleamines*) serve almost exclusively as neuromodulators. Catecholamines are derived from the amino acid *tyrosine*, and three catecholamines—*dopamine*, *norepinephrine*, and *epinephrine*—are important in brain function. Indoleamines are derived from the amino acid *tryptophan*, and one indoleamine, *serotonin*, is a key substance released at brain synapses. *Figure 18* presents the structure of dopamine and serotonin and the amino acids from which they derive.

Monoamines are critical in regulating the brain's mood, i.e., its affective and arousal states. Drugs that alter transmission at synapses using one of the monoamines, or that alter the levels of these substances at synaptic sites, often dramatically change a person's mood or other mental state. The hallucinogen lysergic acid diethylamide (LSD) is a good example, because it interferes with membrane receptors activated by serotonin, thereby preventing serotonin from interacting with the receptors. The amphetamines, another example, are brain stimulants that induce hyperactivity and inability to sleep. They increase levels of catecholamines at synapses, particularly dopamine, by releasing these substances from nerve terminals and by interfering with the reuptake of catecholamines into the presynaptic terminals.

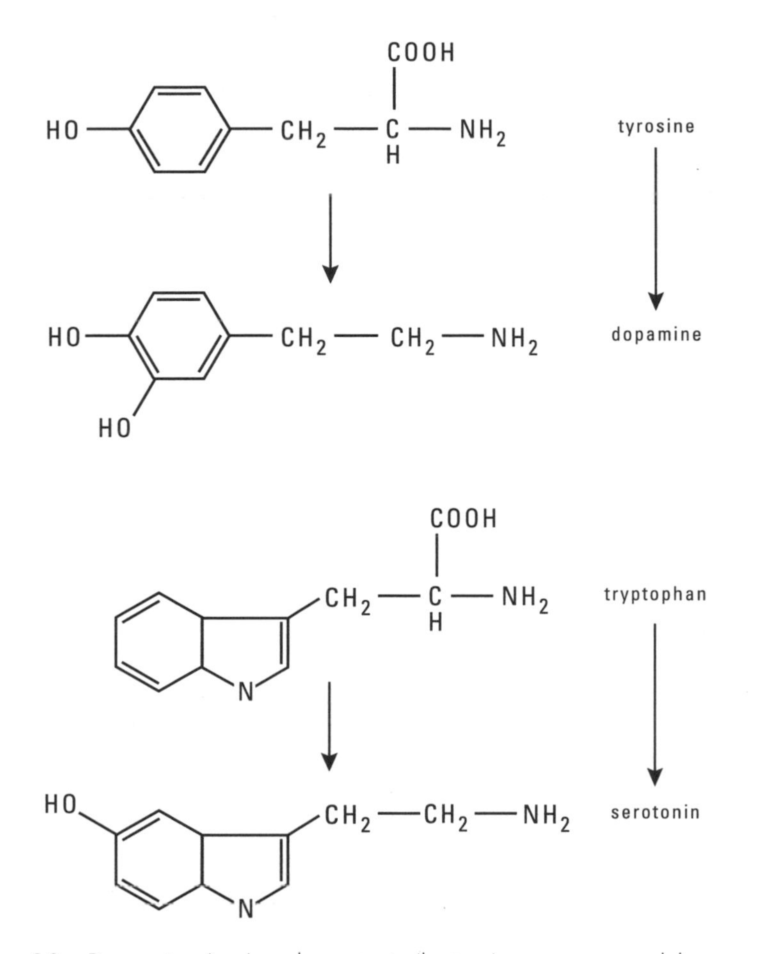

Figure 18 Dopamine (top) and serotonin (bottom), two neuromodulatory substances, are derived from the amino acids tyrosine and tryptophan.

Dopamine, Parkinson's Disease, and Schizophrenia

The monoamines' range of activities is broad and by no means entirely known. For example, two quite different diseases of the brain, Parkinson's disease and schizophrenia, have been related to altered dopamine levels. Parkinson's disease is primarily a disease of the motor system. Patients suffering from Parkinson's disease develop a severe tremor and have trouble initiating movements. They also become rigid and their movements are characteristically slow.

In the late 1950s, it was discovered that the brain's dopamine content was low in patients suffering from Parkinson's disease. Much of the brain's dopamine is in nucleii, termed the *basal ganglia*, that are involved in the initiation of movement. In one basal ganglia nucleus, more than 90 percent of the dopamine may be lost in the disease.

In the early 1960s, a drug therapy for Parkinson's disease was introduced that raises the brain's dopamine levels. The drug, L-dopa, a precursor of dopamine, is effective because it passes readily from the blood to the brain. Dopamine itself does not do that; it is blocked by the blood-brain barrier, which excludes many substances (good *and* bad) from entering the brain. Once in the brain, L-dopa is converted to dopamine and increases the brain's levels of the drug. Although L-dopa causes side effects and sometimes induces schizophrenia-like symptoms, it is successful in many patients and significantly reduces symptoms of Parkinson's disease.

In recent years, another therapy for Parkinson's disease has been tried, with mixed results. With this therapy, cells are transplanted into the brain to release dopamine. Such transplants are expected to raise dopamine levels. Though neural and nonneural cells have been tried, the best results so far have been with fetal cells derived from basal ganglia nucleii. Although the results are promising, cell transplantation is not yet as effective as drug therapy.

Schizophrenia is a severe mental disorder is which patients' judgment is impaired and they lose contact with reality. Schizophrenics' symptoms include thought disorders, in which thoughts and speech are disconnected; delusions and hallucinations; mood disorders, including depression, anxiety, and euphoria; restlessness or inactivity; and withdrawal from social interactions. Two observations suggest that schizophrenia may be due to alterations in dopamine synaptic transmission.

First, the amphetamines that raise brain dopamine levels can induce in humans a state that closely resembles schizophrenia. Second, the standard treatment for schizophrenia is drugs that block dopamine receptors; these drugs prevent dopamine from activating dopamine receptors. The most commonly prescribed agents are haloperidol and clozapine, and they are quite effective in treating many symptoms, especially delusions and hallucinations. For many schizophrenics these antipsychotic drugs have been a godsend because they enable them to function reasonably well in society.

Although indirect evidence suggests that the underlying cause of schizophrenia involves dopamine synapses, we really have no direct evidence. As yet, no study has provided unequivocal evidence for altered dopamine neurotransmission in the brains of patients suffering from schizophrenia. No region of the brain has been found, for example, that possesses excess levels of dopamine.

A problem with treating any brain disease with drugs is side effects. Schizophrenics are given drugs that block dopamine action; thus the presumption is that their brains contain excess dopamine. Parkinson's disease, in contrast, is due to a deficit of dopamine. Do schizophrenics treated with dopamine-blocking agents develop Parkinson-like symptoms? They often do, which limits the amounts and types of dopamine-blocking agents that can be given to schizophrenics. Conversely, some people with Parkinson's disease develop schizophrenic-like symptoms when given L-dopa, presumably because dopamine levels are increased in some region of the brain. For certain patients, the schizophrenic-like symptoms are so severe that L-dopa therapy must be discontinued. Withdrawing the L-dopa therapy reduces the schizophrenia-like symptoms, but then Parkinsonian symptoms reappear.

Serotonin and Depression

All of us have experienced a bout with depression. For some people, depression can become prolonged and debilitating, as was the case for Tess and her mother described at the beginning of the chapter. The possibility that depression may be related to an upset in monoamine synaptic transmission came from observations in the 1950s, when patients were treated with two kinds of drugs for reasons totally unrelated to brain function. Reserpine, which treats hypertension, caused depression in many individuals. It was found subsequently that reserpine depletes monoamines in nerve terminals, and so it was supposed that depression may relate to depressed levels of monoamines. By contrast, drugs for treating tuberculosis (iproniazid and isoniazid) often alleviated depression. These antitubercular drugs are inhibitors of an enzyme, monoamine oxidase, that breaks down monoamines in the brain. In patients treated with iproniazid or isoniazid, therefore, monoamine levels are elevated.

These early observations led pharmacologists to develop and test several monoamine oxidase inhibitors as possible antidepressants, and many of them did show antidepressant activity. But side effects and low efficacy in patients with severe depression led pharmacologists to seek drugs that raise monoamine levels in a different way, by inhibiting the reuptake of monoamines into nerve terminals. The resulting compounds, called *tricyclics*, are more successful in treating depression than are monoamine inhibitors. Like the monoamine inhibitors, however, the early tricyclic drugs were not very specific with regard to a monoamine; all the monoamines were affected, as were other neuroactive substances, and this often led to unpleasant side effects.

The more effective tricyclics inhibited the uptake of just two monoamines, serotonin and norepinephrine, and so a third generation of antidepressant drugs was developed to be more specific in the monoamine reuptake they inhibit. One of these, fluoxetine (Prozac), potently and selectively inhibits serotonin uptake by inhibiting the transporter that is present at synapses releasing serotonin. Thus, it specifically increases serotonin levels at these synapses. This drug has been wildly successful; it is effective with all kinds of people and with a number of affective disorders aside from depression. The drug has few side effects and is tolerated well by most patients. It was the drug that caused the amazing transformation of Tess, described at the beginning of this chapter.

The lesson here is that the more specific a drug is in relation to the molecules at a synapse it affects, the more successful it may be in treating a disorder and the fewer side effects it may have. For example, the receptors that interact with a specific neuromodulator (or neurotransmitter) are not all identical. Typically there are several subtypes of receptors that differ slightly in their protein structure. At any one synapse, usually only one receptor subtype is present. Receptor subtypes can be distinguished pharmacologically, and one challenge for pharmacologists is to find compounds that interact with a specific subtype of receptor. The new drugs that combat schizophrenia primarily block the effect of dopamine on only one of the four known dopamine receptors. These newer drugs, such as clozapine, cause fewer Parkinsonian side effects than the less specific drugs such as haloperidol.

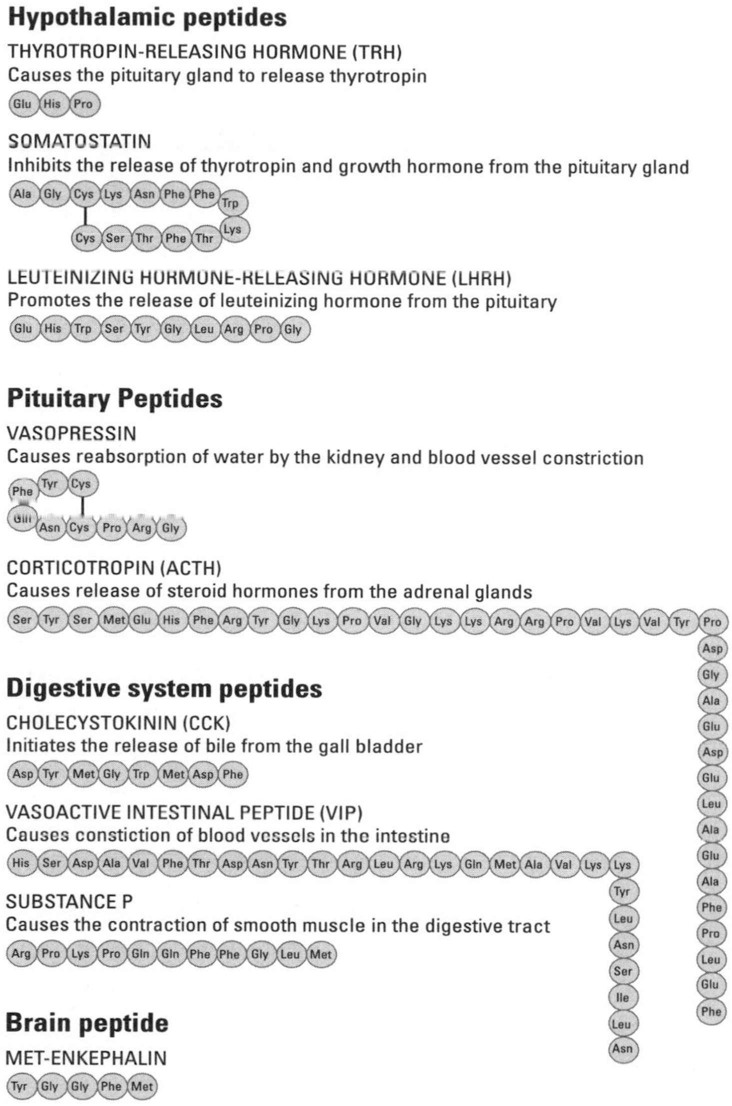

Hypothalamic peptides

THYROTROPIN-RELEASING HORMONE (TRH)
Causes the pituitary gland to release thyrotropin

Glu His Pro

SOMATOSTATIN
Inhibits the release of thyrotropin and growth hormone from the pituitary gland

Ala Gly Cys Lys Asn Phe Phe Trp
Cys Ser Thr Phe Thr Lys

LEUTEINIZING HORMONE-RELEASING HORMONE (LHRH)
Promotes the release of leuteinizing hormone from the pituitary

Glu His Trp Ser Tyr Gly Leu Arg Pro Gly

Pituitary Peptides

VASOPRESSIN
Causes reabsorption of water by the kidney and blood vessel constriction

Phe Tyr Cys
Gln Asn Cys Pro Arg Gly

CORTICOTROPIN (ACTH)
Causes release of steroid hormones from the adrenal glands

Ser Tyr Ser Met Glu His Phe Arg Tyr Gly Lys Pro Val Gly Lys Lys Arg Arg Pro Val Lys Val Tyr Pro Asp Gly Ala Glu Asp Glu Leu Ala Glu Ala Phe Pro Leu Glu Phe

Digestive system peptides

CHOLECYSTOKININ (CCK)
Initiates the release of bile from the gall bladder

Asp Tyr Met Gly Trp Met Asp Phe

VASOACTIVE INTESTINAL PEPTIDE (VIP)
Causes constiction of blood vessels in the intestine

His Ser Asp Ala Val Phe Thr Asp Asn Tyr Thr Arg Leu Arg Lys Gln Met Ala Val Lys Lys Tyr Leu Asn Ser Ile Leu Asn

SUBSTANCE P
Causes the contraction of smooth muscle in the digestive tract

Arg Pro Lys Pro Gln Gln Phe Phe Gly Leu Met

Brain peptide

MET-ENKEPHALIN

Tyr Gly Gly Phe Met

Figure 19 Representative neuropeptides found in the brain. Many of these neuropeptides are involved in endocrine function and were first identified in the hypothalamus, pituitary gland, or digestive system. Their roles in endocrine function are noted here along with their amino acid structures. The amino acid abbreviations are: Ala, alanine; Arg, arginine; Asn, asparagine; Asp, aspartic acid; Cys, cysteine; Gln, glutamine; Glu, glutamic acid; Gly, glycine; His, histidine; Ile, isoleucine; Leu, leucine; Lys, lysine; Met, methionine; Phe, phenylalanine; Pro, proline; Ser, serine; Thr, threonine; Trp, tryptophan; Tyr, tyrosine; Val, valine.

Neuropeptides: The Enkephalins—Endogenous Opiates

By far the most diverse group of substances released at synapses are the neuropeptides. They are also the least understood in terms of their mode of action, but it is believed that neuropeptides generally act as neuromodulators. Alterations in brain levels of several neuropeptides have been linked to mood disorders. Some of these peptides interact with the dopaminergic and serotonergic systems in the brain, and the mood disorders may be accounted for by their effects on monoamine synaptic transmission. It may also be the case that certain peptides themselves play a central role in affective states of the brain.

As many as thirty different peptides are released at synapses. These peptides vary widely in size, containing between three and forty amino acids. They are commonly divided into four groups, depending on where they were first identified. For example, two groups, the hypothalamic peptides and pituitary peptides, participate in the release of hormones from the pituitary or other glands. A third group was identified first in the digestive system, where it is responsible for regulating aspects of digestion. Finally, some peptides were first identified in brain tissue. *Figure 19* lists groupings and representative examples.

In any region of the brain the number of neurons containing and releasing a peptide from its terminals is small. The peptide-containing neurons, though, trigger wide-ranging processes, which means that their role is a more global one. But as yet, specific roles for most peptides in brain function have not been established. A striking exception, though, is the enkephalins, which appear to act as natural opiates. Opiates such as morphine, the active ingredient of the opium poppy, have long been used to relieve pain and to treat medical problems ranging from coughing to diarrhea. The opiates also cause pleasant mental effects and are classic recreational drugs. It is not news that the opiates are highly addicting and have presented a serious problem around the world for centuries. After the Opium Wars of the early nineteenth century, when the British forced open the Chinese market and permitted the massive importation of opium into China, a fourth of the Chinese population became addicted to opium. Today, society has the same problem with heroin, a chemically modified and more potent form of morphine.

The discovery of enkephalins was prompted by hints that receptors for opiates might reside within the brain. Why? For two reasons. The first was that many opiates are effective at tiny concentrations, which suggested the possibility of specific recognition sites for such molecules in the brain. The second was the discovery of substances that block the effects of opiates. One such drug, naloxone, seemed to act like other substances that prevent a neurotransmitter or neuromodulator from interacting with its channel or receptor.

In the 1970s, Candace Pert and Solomon Snyder at Johns Hopkins University proved that brain tissue has specific receptors for opiates; these findings suggested that natural opiate-like substances are in the brain. Such substances were looked for in brain tissue, and shortly thereafter, two small peptides that bind to opiate receptors, the enkephalins, were isolated from brain tissue.

What role might the enkephalins play in brain function? The answer remains elusive, but their discovery explains some puzzling aspects of pain and how it can be relieved in unconventional ways. For example, the relief of pain by acupuncture treatment probably results from the release of enkephalins within the brain. Evidence that this is so comes from experiments on animals, which were shown to experience higher pain threshold levels than controls when given acupuncture treatment. However, if the animals were first given naloxone, which presumably blocked the opiate receptors in their brains, acupuncture treatment had no effects on pain thresholds.

The placebo effect that results in pain relief is another phenomenon that may be explained by the endogenous release of opiate-like substances. A convincing experiment was conducted at the University of California San Francisco Medical Center by Howard Fields and his colleagues. Medical students who had had a wisdom tooth extracted were given either morphine or a sugar pill placebo. They were told this therapy would decrease pain, and both groups experienced less pain than a control group given nothing. When a subset of the students given the placebo was also given naloxone, these students experienced pain levels comparable to the controls given nothing. The conclusion drawn was that a pill thought to relieve pain released enkephalin or enkephalin-like substances in the brain.

Of what value would opiates released within the brain be? Soldiers severely wounded in battle and athletes hurt in competition often feel little pain until the battle or contest is over. Why? The release of endogenous opiates could explain the lack of pain. Even strenuous activity appears to release endogenous opiate-like substances, as shown by the block by naloxone of the "high" often experienced by runners after a long run. One can envision the survival or merciful value of such substances released in the brains of wounded humans or animals that are being attacked or chased. A vivid description of how powerful this system may be is provided by David Livingstone, a Scottish missionary and explorer who was attacked by a wounded lion in Africa. Fortunately for Livingstone, the lion was quickly distracted by another member of his party who was attempting to shoot the beast. The lion left Livingstone to attack the gunman. A third man was also bitten before the lion fell dead from his wounds. All three survived the lion's attacks. Livingstone wrote of his reactions to the attack in his *Missionary Travels* (John Murray, London, 1875):

> . . . I heard a shout. Starting, and looking half around, I saw the lion just in the act of springing upon me. I was upon a little height, he caught my shoulder as he sprung, and we both came to the ground below together. Growling horribly close to my ear, he shook me as a terrier does a rat. The shock produced a stupor similar to that which seems to be felt by a mouse after the first shake of the cat. It caused a sort of dreaminess in which there was no sense of pain nor feeling of terror, though quite conscious of all that was happening. It was like what patients partially under the influence of chloroform describe, who see all the operation, but feel not the knife. . . . The shake annihilated fear, and allowed no sense of horror in looking round at the beast. This peculiar state is probably produced in all animals killed by the carnivora; and if so, is a merciful provision by our benevolent creator for lessening the pain of death.

Other receptors and endogenous substances may mediate similar types of phenomena. For example, a receptor for marijuana and marijuana-like substances has been identified in brain tissue, and endogenous substances that bind to these receptors have been found. What role such receptors or substances may play is not known.

To conclude, despite all that is still mysterious about brain chemistry, drug therapy for mental disease has been a tremendous success story. Many severely disturbed patients previously doomed to a life in a mental institution can now function in society as a result of drug therapy, and less debilitating mental disturbances such as anxiety and depression can often be relieved quite effectively with a variety of substances. Drug therapy is by no means a panacea; drugs treat symptoms and don't cure the disease. For many, side effects are a problem, and in some cases, drug therapy fails. Nevertheless, drug therapy has been an enormous advance for the treatment of mental disease, and as we learn more of the intricacies of brain function, especially synaptic mechanisms, new, more effective and specific agents will undoubtedly be found.

Prior to drug therapy, the most effective treatment for mental illness was, of course, psychotherapy. But psychotherapy was not particularly effective with severely ill patients, and it is slow and expensive. The great success with drug therapy naturally raises the question of how psychotherapy affects the brain. There is evidence that psychotherapy can alter brain chemistry, and so the two treatments may be more complementary than is intuitively obvious. How might psychotherapy alter brain chemistry? The placebo phenomenon described earlier may provide a model. That is, if a subject believes a pill will relieve pain, it does so even if the pill is pure sugar. Ingestion of the placebo causes the release of endogenous opiate-like substances in the brain—brain chemistry is altered and profound effects can result. It is quite conceivable that a similar type of alteration in brain chemistry—an increase or decrease in the release of a neurotransmitter or neuromodulator—occurs as a result of psychotherapy. Indeed, most psychiatrists today view psychotherapy as an essential adjunct of drug therapy. They believe drug therapy is more effective when combined with psychotherapy.

CHAPTER 4

Creepy, Crawly Nervous Systems

The Marine Biological Laboratory (MBL) in Woods Hole is a paradigm, a human institution possessed of a life of its own, self-regenerating, touched all around by human meddle but constantly improved, embellished by it. . . .

Successive generations of people in bunches, never seeming very well organized, have been building the MBL since it was chartered in 1888. It actually started earlier, in 1871, when Woods Hole, Massachusetts, was selected for a Bureau of Fisheries Station and the news got round that all sorts of marine and estuarine life could be found here in the collisions between the Gulf Stream and northern currents offshore, plus birds to watch. Academic types drifted down from Boston, looked around, began explaining things to each other, and the place was off and running. . . .

The invertebrate eye was invented into an optical instrument at the MBL, opening the way to modern visual physiology. The giant axon of the Woods Hole squid became the apparatus for the creation of today's astonishing neurobiology. Developmental and reproductive biology were recognized and defined as sciences here, beginning with sea-urchin eggs and working up. Marine models were essential in the early days of research on muscle structure and function, and research on muscle has become a major preoccupation at the MBL. Ecology was a sober, industrious science here long ago, decades before the rest of us discovered the term. In recent years there have been expansion and strengthening in new fields; biologic membranes, immunology, genetics, and cell regulatory mechanisms are currently booming.

You can never tell when new things may be starting up from improbable lines of work. The amebocytes of starfish were recently found to contain a material that immobilizes the macrophages of mammals, resembling a product of immune lymphocytes in higher forms. *Aplysia*, a sea slug that looks as though it couldn't be good for anything, has been found by neurophysiologists to be filled with truth. *Limulus*, one of the world's conservative beasts, has recently been in the newspapers; it was discovered to contain a reagent for the detection of vanishingly small quantities of endotoxin from gram-negative bacteria, and the pharmaceutical industry has already sniffed commercial possibilities for the monitoring of pyrogen-free materials; horseshoe crabs may soon be as marketable as lobsters.

—Excerpted from Lewis Thomas, *The Lives of a Cell: Notes of a Biology Watcher* (New York: Penguin USA, 1995)

The nervous systems of animals without backbones—invertebrates—are simpler, with fewer neurons, than those found in cats or catfish, frogs and finches, the vertebrates. Because of the relative simplicity of invertebrate nervous systems, neuroscientists have long studied them, and they have reached some truly wonderful insights about brain function. The underlying assumption is that the same biological mechanisms operate in the nervous systems of all organisms—a not incorrect view. Invertebrates have provided invaluable knowledge about how single nerve cells function and how neuronal groups give rise to simple behaviors. Some of this research has revealed clues to higher brain phenomena like perception, learning, and memory.

The sea is a treasure-house of especially useful and accessible organisms. The squid, with its giant nerve cell axon, has provided much of what we know about how nerve cells generate electrical signals. The horseshoe crab has an optic nerve that easily frays, and so it was first possible to record single optic nerve axons in that animal. Subsequent studies of the horseshoe crab visual system provided insights into how we perceive edges and borders. The sea slug Aplysia *has a robust gill withdrawal reflex, one that is highly modifiable. This animal has elucidated how nerve cell function can be altered by experience—how a nervous system can learn and remember things. And much of this work was initiated and carried out at the MBL in Woods Hole.*

Electrical Signaling and the Squid Giant Axon

In vertebrates, as was described earlier, most axons are covered by an insulting layer of myelin that accelerates action potential generation down the axon and streamlines its propagation. The glial cells of invertebrates, except in a few crabs, do not form myelin around nerve cell axons—invertebrate axons are without the benefit of myelin. An obvious consequence of the absence of myelin around invertebrate axons is that they conduct action potentials much slower than do vertebrate axons. While action potentials typically travel down vertebrate axons at rates of 100–200 miles per hour, they move along invertebrate axons at no more than 30–40 miles per hour. But even myelinated axons conduct action potentials incredibly slowly compared with how fast electrons flow down a conducting wire—at the speed of light!

Invertebrates can maximize how fast the action potential conducts down axons by making their axons larger. When they do so, it lowers the axon's internal electrical resistance—that determines how easily ions can flow down the axon. Myelin, on the other hand, increases the cell membrane's electrical resistance—that governs how easily ions can flow across the membrane. What is critical for a high rate of action potential conduction down an axon is the ratio of membrane resistance to internal resistance. The higher the ratio, the faster is action potential conductance. So, decreasing an axon's internal resistance or increasing an axon's membrane resistance does very much the same thing biophysically; it speeds conduction. But enlarging axon size to raise its conduction speed has limitations. Axons can be made only so big, and even the largest invertebrate axons conduct action potentials slower than the great majority of myelinated vertebrate axons. Some say that if we did not form myelin, our brains would have to be ten times larger and we would need to consume ten times more food to maintain a nervous system comparable to the one we have. And one reason why invertebrates have not developed more complex nervous systems is that their glial cells do not form myelin around axons—they are limited in the number of fast-conducting axons in their nervous systems.

In squid, huge giant axons have evolved to mediate escape responses (*Figure 20*). The giant axons in squid may be as large in diameter as 1 mm. Indeed, the axons are so large that they were once thought to be blood vessels running in squid nerve axon bundles. In the mid-1930s, an English scientist, John (J. Z.) Young, who was spending the summer at the MBL, realized that the long giant axon running down the length of the animal contained no blood cells and that it appeared similar histologically to the smaller axons surrounding it. He proposed that it was a nerve, and with physiologists at the MBL, including H. K. Hartline, he proved that it was indeed a giant nerve cell axon. When this giant axon generates action potentials, powerful muscles throughout the squid's body contract and squirt water from one end of the animal, enabling it to escape by jet propulsion. These muscles have to activate quickly—hence the giant axons.

Once the giant axon was identified as a nerve, it underwent intense investigation. In the late 1930s, scientists showed that by squeezing out

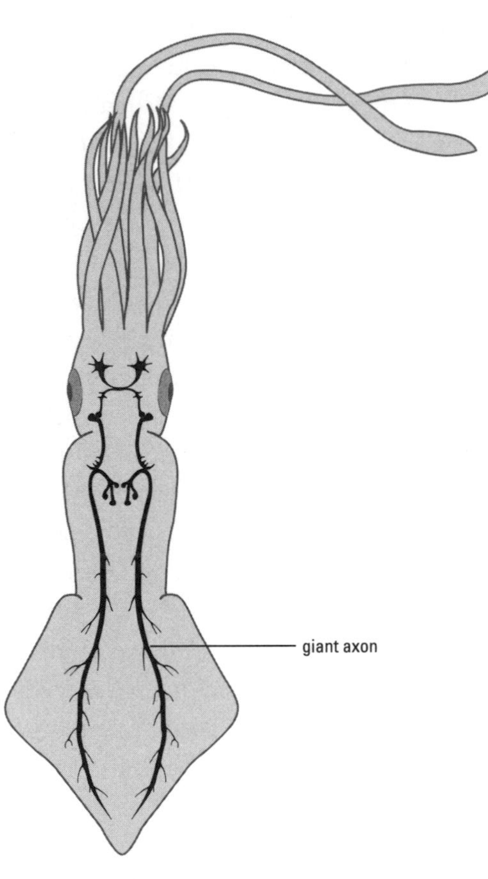

Figure 20 The giant axon system of the squid. Three pairs of giant axons exist: the brain directly innervates the first giant axons (top), which in turn activate the second giant axons (middle). The second giant axons synapse on the third giant axons, which run the length of the squid's body. The third-order giant axons, formed by the fusion of axons from a number of neurons, are those used for most physiological experiments.

the contents of a giant axon, the ionic and other constituents inside cells could be determined: K^+ levels are high inside the axon and Na^+ and Cl^- concentrations are low. Next physiologists inserted electrodes inside the axon and measured the voltage across the membrane when it was at rest—the resting potential—and when an action potential was generated. These measurements clarified that K^+ is the key ion for establishing the resting potential and Na^+ the key ion for action potential generation. Soon it became possible to free the nerve from the surrounding axons and experiment on an isolated axon, as well as to perfuse the axon with artificial solutions and manipulate inside concentrations of ions or other constituents.

Much of this work was done by two English scientists, Alan Hodgkin and Andrew Huxley, at the Marine Biological Station in Plymouth, England, in the late 1940s and early 1950s. Hodgkin first experimented on the squid giant axon at the MBL in the summer of 1938, working there with Kenneth (K. C.) Cole, an American scientist who pioneered physiological techniques for analyzing the giant axon's response. In 1952, Hodgkin and Huxley provided an analysis of action and resting potentials of the squid giant axon that holds to this day and is applicable to the nerve cell axons of all animals, including humans. For this work Hodgkin and Huxley were awarded the Nobel Prize in 1963.

The squid giant axon remains the largest single adult cell known, and so it continues to be of great experimental value. Each summer, scientists flock to Woods Hole or other marine stations around the world to experiment on it. Now the questions have become: How does the voltage change across the axonal membrane open Na^+ channels, and what are the mechanisms for transporting substances down the inside of an axon? If substances move down axons only by passive diffusion, it would take fifteen years to traverse the 1-meter-long axon of the spinal cord motor neuron described in Chapter 1! Axons, therefore, have devised special transport mechanisms that speed substances along their interiors. This process of *axonal transport* is of particular importance because proteins cannot be made in the axons or axon terminals of neurons, for reasons that are not understood; axonal structures depend on proteins made in the neuron's cell body for their maintenance. If axonal transport is disrupted, axonal and terminal function soon fails. Thus the way axonal transport works is of considerable interest, and we can visualize it in squid axons by using video-enhanced microscopy techniques, a technique pioneered at the MBL.

Mach Bands and the Horseshoe Crab Eye

Another exceptionally useful marine invertebrate "discovered" at the MBL is the horseshoe crab, *Limulus polyphemus*. *Limulus* is technically not a crab but a member of the spider family. Like many other creatures in the arthropod (insect) phylum, the horseshoe crab has two prominent faceted eyes. The facets consist of photoreceptive units called *ommatidia*, and each eye in *Limulus* has about twelve hundred of these units. Each ommatidium is made up of about fifteen cells sensitive to light (photoreceptor cells) and one second-order neuron, called an *eccentric cell*, which collects information from the photoreceptor cells. The photoreceptor cells are linked to the eccentric cell by electrical synapses.

When a photon of light is caught by one of the photoreceptor cells, channels open in the cell's membrane that allow Na^+ to enter, thus depolarizing the cell. This response is a receptor potential, which, via the electrical synapse, passes directly into the eccentric cells. Extending from the eccentric cell is an axon that, if sufficiently depolarized, will fire action potentials, which then travel down the axon. The optic nerve of *Limulus*, extending from the eye to brain, consists primarily of eccentric cell axons. A drawing of a horseshoe crab and its faceted eye and a diagram of a longitudinal section through an ommatidium are shown in *Figure 21*.

How many action potentials are generated by an eccentric cell axon reflects to a first approximation the number of light quanta caught by the photoreceptors. But the real-life situation is more complicated, and this discovery led to an elegant explanation of a psychological phenomenon, *Mach bands*, seen at the border between two areas illuminated by different light intensities.

Why was *Limulus* chosen for visual studies? As a college student, H. Keffer Hartline had become interested in visually guided behavior when he found that pill bugs avoid light. His undergraduate research project analyzed this behavior, and in the late 1920s, he spent a summer at Woods Hole examining several invertebrates in search of a visual system from which he could record the activity of single axons coming from the eye. What attracted him to the horseshoe crab was its long optic nerve, which runs just underneath the shell from the eye to the anteriorly situated brain (*see Figure 21*). Another attraction is that it is so easy to dissect the nerve. In most nerve bundles, connective tissue binds axons together so tightly it is difficult to tease apart individual fibers. Not so in *Limulus*. Isolating single optic nerve fibers by dis-

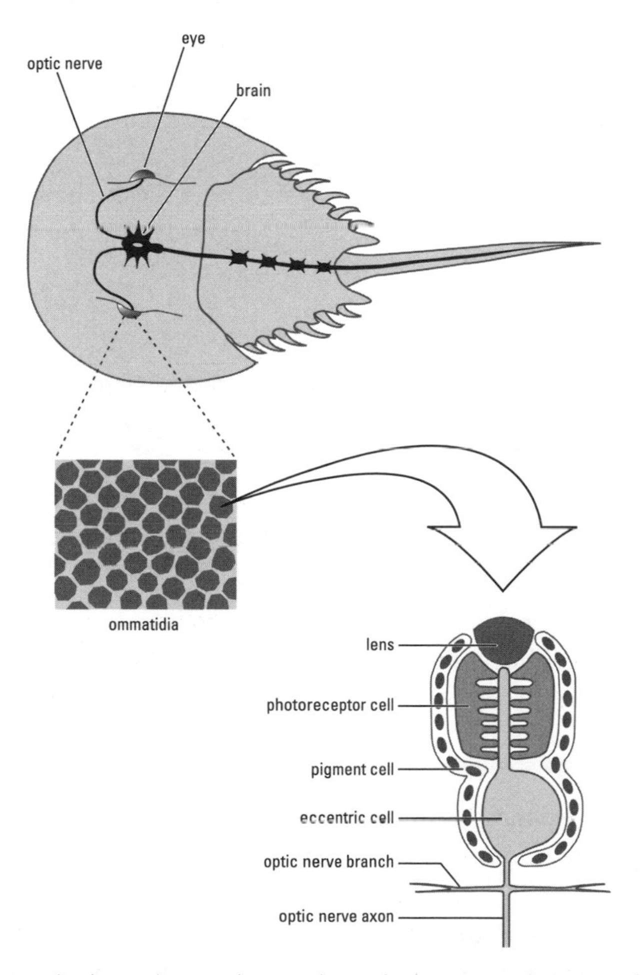

Figure 21 The horseshoe crab, *Limulus polyphemus*, and its eye. The optic nerve running from the eye to the brain is located just under the shell, making it highly accessible. The eye is made up of individual units, called ommatidia, that consist of a lens, photoreceptor cells, pigment cells, and an eccentric cell from which the optic nerve axon arises.

section is straightforward, as generations of students can attest (*see Figure 22*). Recording from the optic nerve of *Limulus* has been a favorite laboratory exercise in biology and neurobiology courses.

Hartline's recordings were the first single cell recordings made from a visual system. He first examined the responses of individual axons to light stimuli and showed that the response paralleled many features of the human visual system. For example, if he used short flashes (less than one second), the response he recorded depended strictly on how many photons impinged on the eye. Thus, reciprocal changes in intensity and duration elicited the same response in an axon so long as the flash had the same number of photons. This phenomenon, known as Bloch's law, had long been known in humans; a short, intense flash looks identical to a longer, dimmer flash if both flashes have the same number of photons. These data, then, suggested that the horseshoe crab eye could elucidate the human eye's function.

When Hartline began his work, and for many years thereafter, he thought that the ommatidia were completely independent units—that they did not interact. One day, as he was recording from an optic nerve fiber and stimulating only the ommatidium where the optic nerve fiber came from, he noticed that when he turned on the room lights, the recorded axon decreased its activity. Since more light was falling on the eye when the room lights were on, this observation was puzzling. Adding more light to the eye should enhance activity, not dampen it. One explanation was that surrounding ommatidia, now illuminated by the light, inhibited the ommatidium whose activity was being recorded, and this turned out to be true. Hartline and his colleague Floyd Ratliff proved that illuminating surrounding ommatidia could reduce activity in the recorded ommatidium. They called this *lateral inhibition*, and we now know that it occurs in all visual systems.

What purpose does this phenomenon serve for visual processing? Hartline and Ratliff's extensive analyses of lateral inhibition in the horseshoe crab eye in the late 1950s and 1960s demonstrated this elegantly. The inhibitory effects are mediated by fine branches of the eccentric cell axons that arise shortly after the axons emerge from the ommatidia (*Figure 21*). The branches extend laterally to form inhibitory synapses on adjacent eccentric cell axons. The inhibition is reciprocal; an axon inhibits its neighbor, which in turn is inhibited by that neighbor.

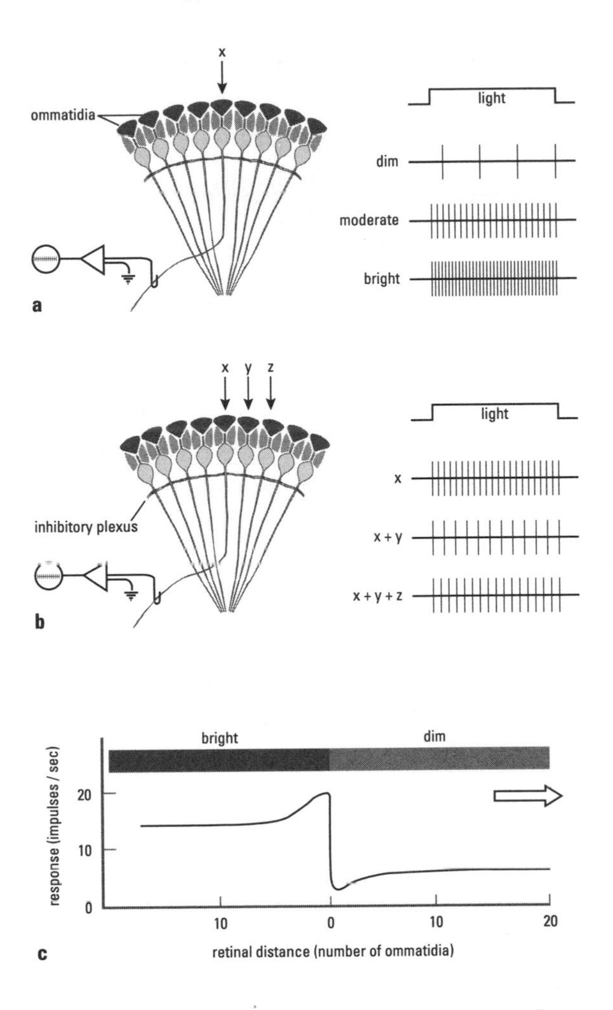

Figure 22 Recording from optic nerve axons in *Limulus*. *a*: Illuminating a single ommatidium (X) and recording from the optic nerve axon coming from that ommatidium results in activity levels that depend strictly on light intensity. Dim light evokes a weak response, bright light a vigorous response. *b*: Recording from a single optic nerve axon but illuminating more than one ommatidium results in activity whose level depends not only on light intensity but also on lateral inhibitory effects mediated by the inhibitory plexis. Illuminating X + Y simultaneously reduces the activity generated when only X is illuminated because of the lateral inhibition exerted by Y on X. Illuminating Z in addition to X and Y causes inhibition of Y, partially relieving the inhibition of Y on X. Hence, activity in the optic nerve coming from X increases. *c*: The interplay of excitatory and inhibitory effects cause an accentuation of contrast at a border between bright and dim lights. Activity is greater along the bright side of the border, and lower along the dim side of the border, than is activity some distance away from the border.

The effects of excitation and lateral inhibition on the response of an eccentric cell axon are depicted in *Figure 22*. The axon coming from ommatidium X is recorded with a simple wire hook electrode. If ommatidium X alone is illuminated, the number of action potentials generated per unit time depends on the intensity of light falling on the ommatidium (*Figure 22a*). A dim light evokes few action potentials; a bright light many more. On the other hand if the light intensity is kept constant and of moderate intensity, illuminating ommatidium Y decreases the firing frequency of the axon from ommatidium X (*Figure 22b*). This is the effect of lateral inhibition. The strength of lateral inhibition between eccentric cell axons depends on how activated the interacting units are and the distance between them; the stronger the illumination, the stronger the inhibition exerted—and near neighbors affect one another more than do distant ones. A further consequence of such a laterally inhibiting network is demonstrated when a third ommatidium, Z, is illuminated. Illuminating Z inhibits Y, which decreases the inhibition of X by Y—a phenomenon called *disinhibition*.

What significance does this system of lateral, reciprocal, and disinhibition have for vision? Hartline and Ratliff proved convincingly that such interactions can enhance contrast at an edge or border (*Figure 22c*). If a step of light is projected onto an array of ommatidia, the axons coming from the eccentric cells are more active along the bright side of the bright-dim edge than are axons away from the edge; conversely, axons adjacent to the dim side of the edge are less active than axons farther away. This happens because an eccentric cell axon in bright light that is adjacent to a dark border is inhibited weakly by its neighbors in the dimmer light and, conversely, axons along the dimmer side are inhibited strongly by neighbors over the border in bright light. In other words, axons adjacent to a border are inhibited less or more strongly than are axons away from the border. Because of this interplay between excitatory and inhibitory effects, the differences in activity for axons adjacent to the bright-dim border are greater than they are for axons away from the border. Lateral inhibition shapes the signals coming from the axons in such fashion that the intensity differences across the border are enhanced.

A similar enhancement of borders in the human visual system was recognized a century ago by Ernst Mach, an Austrian physicist and

Figure 23 The Mach band phenomenon. Although the reflected intensity of each square is constant from one border to the next, it appears as if each one is lighter along the border with a darker square and darker along the border of a lighter square. The Mach band phenomenon also accentuates the intensity differences between the steps of the series. If the border between two steps is obscured by a pencil or other thin object, the adjacent steps will appear much more similar in intensity than is the case when the same steps meet at a distinct border.

psychologist. Called the *Mach band phenomenon*, it consists of light steps of increasing intensity from left to right (*Figure 23*). Although each step is even in intensity from one edge to the next, the steps appear lighter along the border adjacent to the darker steps and darker along the borders adjacent to the lighter steps. In the human and other vertebrate eyes, Mach band phenomena result from lateral and reciprocal interactions that occur distally in the retina. The anatomy of the vertebrate eye is different from that of the *Limulus* eye, but the same basic physiological phenomenon is observed; distant photoreceptors inhibit the output of central photoreceptors. In the vertebrate retina, the lateral inhibition is mediated by a separate cell, the horizontal cell.

Learning, Memory, and a Sea Snail

A more recent example of an invertebrate that has provided powerful insights into neurobiological mechanisms comes from the experiments of Eric Kandel, James Schwartz, and coworkers, first at New York University and later at Columbia University's College of Physicians and Surgeons. In research carried out on the sea snail, *Aplysia californica*, they analyzed two elementary forms of learning and memory, habituation and sensitization. Their findings have had a profound impact on our thinking about mechanisms underlying learning and memory.

Aplysia is a rather unattractive animal; it is about 10 inches long and abounds in tide pools along warm-water coasts. It is a soft animal, not enclosed in a shell, that floats near the water surface and eats seaweed. When disturbed, it secretes jets of dark purple ink.

Why were neurobiologists drawn to this marine organism? *Aplysia* presents several important features. First of all, many cells in *Aplysia* are large—0.8–1 mm in diameter—so recording from many of the neurons is relatively easy. The discovery of these large nerve cells in *Aplysia* dates back a century to the early invertebrate histologists. Furthermore, in *Aplysia*, as in many other invertebrates, the nervous system is distributed along the animal rather than collected together in the head. These animals have paired *ganglia*, collections of nerve cells that control one or another behavior. One ganglion might control feeding, another escape behavior, another swimming, and so forth. A ganglion has relatively few neurons, perhaps one thousand to two thousand, so it is possible to record from many of the cells in a single ganglion, identify the cell's function, and explore how the cells are connected together.

When physiologists first began to record from such ganglia in invertebrates, they also found that cells carrying out the same function are located in the same place in the ganglia in animal after animal. Thus, experimenters can construct maps that identify neurons in various ganglia, which greatly facilitate the research. If you want to study a particular cell type, reliable data and diagrams will tell you which ganglion to go to and where in the ganglion the cell will be. Kandel and his colleagues took advantage of this to explore the behavior of, and the neural circuitry underlying, the gill-withdrawal reflex.

Figure 24 shows *Aplysia* from the side and looking down on the animal when the fleshy coverings on its back side, called parapodia, are separated to reveal the gill. Over the gill's base is the mantle, a protective covering, and associated with the mantle is the siphon, a fleshy extension of the mantle. When the mantle or siphon is touched, the gill rapidly withdraws under the mantle to protect itself. Within a few seconds, the gill reappears; but if the mantle or siphon is touched again, the gill disappears again. If the stimuli are continued, the gill continues to withdraw, but with repeated stimuli the gill withdraws less and less. After only four consecutive stimulations spaced one to three minutes apart, gill withdrawal is only about half as extensive as it was originally. And if the stimuli are presented ten times consecutively, at intervals of thirty seconds to a few minutes, gill withdrawal

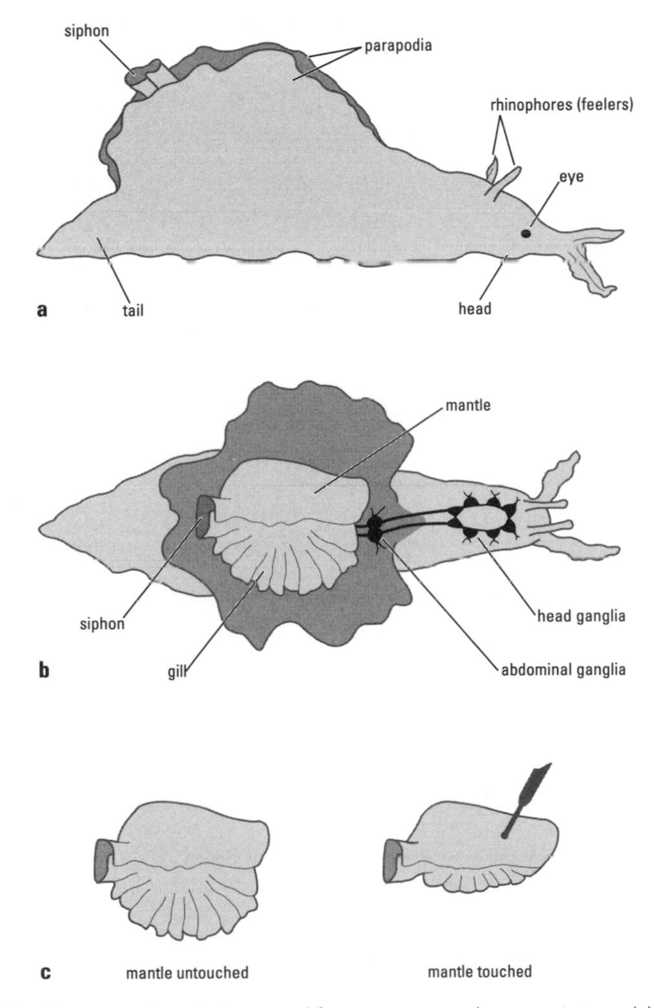

Figure 24 The sea slug *Aplysia californica*, in a side view (*a*) and looking down on the animal from above when the parapodia are spread apart (*b*). The gill, lying underneath the mantle, contracts when the mantle or siphon is touched (*c*). The gill-withdrawal reflex is controlled by neurons in the abdominal ganglia whose approximate location is shown in *b*.

is only about 30 percent what it is at the first stimulus. This decrease in strength after repeated stimulations is called *habituation*.

It is possible to quantify the extent of gill withdrawal simply by placing a photocell under the gill and illuminating the experimental preparation from above. As the gill withdraws, more of the photocell is exposed, which in turn heightens its output. *Figure 25* shows records of habituation of the gill-withdrawal reflex. If the mantle receives ten stimulations and then no others, the response will recover after about two hours.

A second modification of the gill-withdrawal reflex is *sensitization*, which happens when a strong stimulus, such as a sharp pinprick, is applied to the head or tail. In sensitization, gill withdrawal is immediately faster, stronger, and longer than in a nonsensitized animal. Recovery after sensitization of the reflex takes minutes to hours, depending on the extent of sensitization.

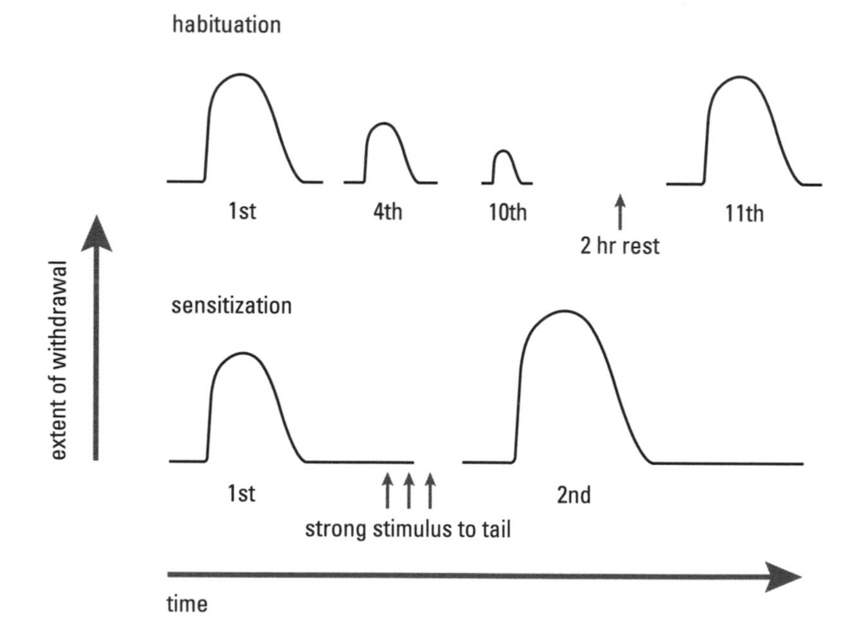

Figure 25 Habituation and sensitization of the gill-withdrawal reflex. Repeated touching of the mantle causes rapid habituation of the withdrawal response; the extent of gill withdrawal decreases significantly. Recovery may take two hours or more (top). A strong stimulus applied to the tail significantly increases the extent of gill withdrawal; it rapidly sensitizes the response (bottom).

An interesting feature of habituation and sensitization in *Aplysia* is that they can be induced to last for weeks if a regimen of training stimuli—say ten consecutive habituating or sensitizing stimuli per day—are given each day over the course of four or five days. Here, either habituation or sensitization can be induced to last for a month or so. These phenomena, then, represent long-lasting changes in a neurally mediated response and thereby provide a model for elementary learning and memory. Furthermore, it is of interest that habituation and sensitization may be either brief (lasting hours) or long (lasting weeks) because of the abundant evidence that we and many other animals have two forms of memory: short-term and long-term. Immediately after we experience something, it is remembered, but these initial memories are quite labile. A blow to the head or intense mental activity may make a person forget a recent event or recently learned material. Individuals in automobile accidents often fail to remember the accident or what happened ten or fifteen minutes prior to it. The survivor of the automobile crash that took the life of Princess Diana showed such an amnesia. He does not remember the crash or events immediately preceding it, but he does remember events of earlier in the evening. Only after a little time (fifteen minutes or so) do memories begin to become stable and more resistant to erasure. One notion is that short-term memories reflect ongoing neural activity whereas long-term memories reflect structural changes in the brain—the formation of new synapses, or the structural alteration of existing synapses.

The gill-withdrawal reflex in *Aplysia* is mediated by one ganglion, the abdominal ganglion, which has about fifteen hundred neurons. By recording sequentially from neuron after neuron, Kandel and his colleagues identified thirty-three neurons in the ganglia involved in the reflex. Twenty-four are sensory neurons; they are stimulated when the mantle or siphon are touched. Six are motor neurons whose activation prompts the gill to withdraw. The motor neurons drive the muscles that effect gill withdrawal. Three are interneurons whose branches are confined to the ganglion. These neurons receive synaptic input from the sensory neurons and synapse on the motor neurons. In addition, the sensory neurons synapse directly on the motor neurons.

Because the reflex is mediated entirely within the abdominal ganglion, which is just anterior to the gill but on the animal's ventral side, it is possible to remove the ganglion from the animal and study the reflex's neural circuitry when the ganglion is maintained in a small dish. The sensory neurons can be activated by electrical pulses to axons that enter the ganglion; the motor neurons' output can be monitored by recording axons leaving the ganglion. Furthermore, the isolated ganglion demonstrates habituation. If incoming sensory axons are repeatedly stimulated, the output of the motor neuron axon subsides, just as the gill's withdrawal lessens when the mantle or siphon is repeatedly touched. *Figure 26* shows a simplified version of the circuitry and the experimental arrangement.

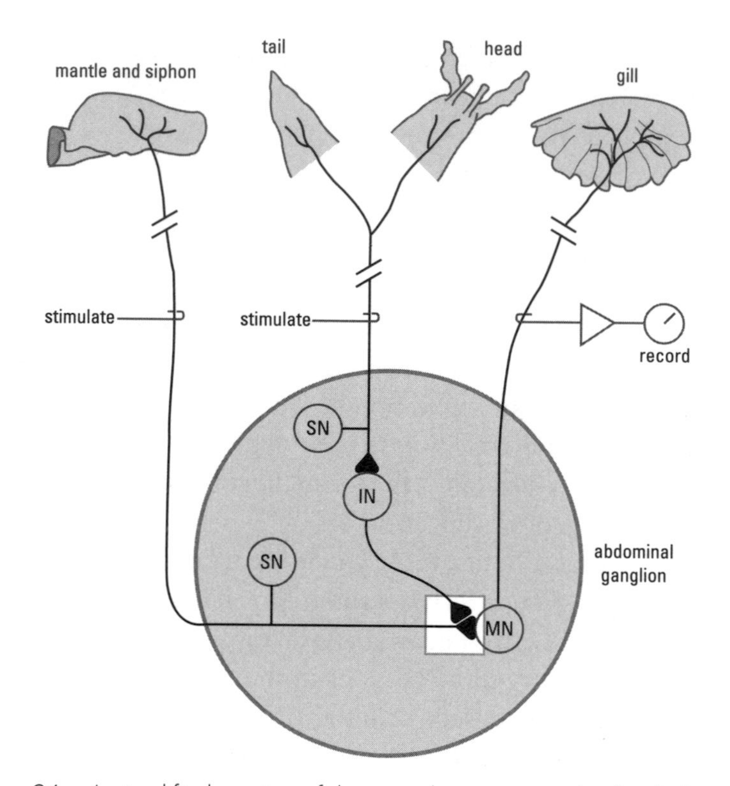

Figure 26 A simplified version of the neural circuitry involved in habituation and sensitization of the gill-withdrawal reflex. Sensory neurons (SN) from the mantle and siphon synapse on motor neurons (MN) that innervate the gill. Sensory neurons from the tail or head synapse on interneurons (IN) that make synapses upon the terminals of the sensory neurons coming from the mantle or siphon (box). The interneurons that are activated by the sensory neurons coming from the mantle and siphon and that synapse directly on the motor neurons are not shown because they play no significant role in habituation.

What, then, causes habituation? Because knocking out the interneurons that synapse on the motor neurons had little effect on habituation, Kandel and colleagues focused on the synapse between the sensory neurons and motor neurons. They discovered that this is an unusual synapse in that it releases less and less neurotransmitter when the sensory neurons are repeatedly stimulated. The decrease in motor neuron output can be accounted for by alterations of the sensory neuron synapses on the motor neurons. And this alteration is a diminished release of transmitter following repeated stimulation of the sensory neurons.

This was a remarkable finding, that a behavioral modification in an animal could be localized to a single set of synapses! What causes the diminished release of neurotransmitter from the sensory terminals? It is still not entirely clear, but Ca^{2+} ions are likely accomplices. Repeatedly stimulating the sensory neurons builds up Ca^{2+} in the terminals, and that may act as a second messenger, binding to calmodulin and activating kinases. The kinases in turn are likely to phosphorylate proteins in the terminal and to diminish neurotransmitter release.

Sensitization also can be induced in isolated abdominal ganglia. Analysis of sensitization revealed similar but even clearer results. To understand how isolated ganglia are sensitized first requires knowing how inputs from the head and tail interact with the gill-withdrawal reflex circuitry. Sensory input from the head and tail impinges on interneurons in the abdominal ganglion that synapse on the sensory neuron terminals. These interneurons make what are called *presynaptic synapses* in which one set of terminals synapse on another set of synaptic terminals. The circuitry is shown in *Figure 26*.

With this knowledge, experimenters stimulated the sensory axons from the head or tail which increased the neurotransmitter release from the sensory neuron terminals. Hence, reflex alterations were again attributed to a change in one set of synapses—indeed, the same set of synapses—and to a change in the amount of neurotransmitter released, but in this case there was an increase of neurotransmitter release, not a decrease.

How this happens was worked out in exquisite detail, which can be summed up very briefly (*Figure 27*). The terminals of the interneurons impinging on the sensory neuron terminals were found to release

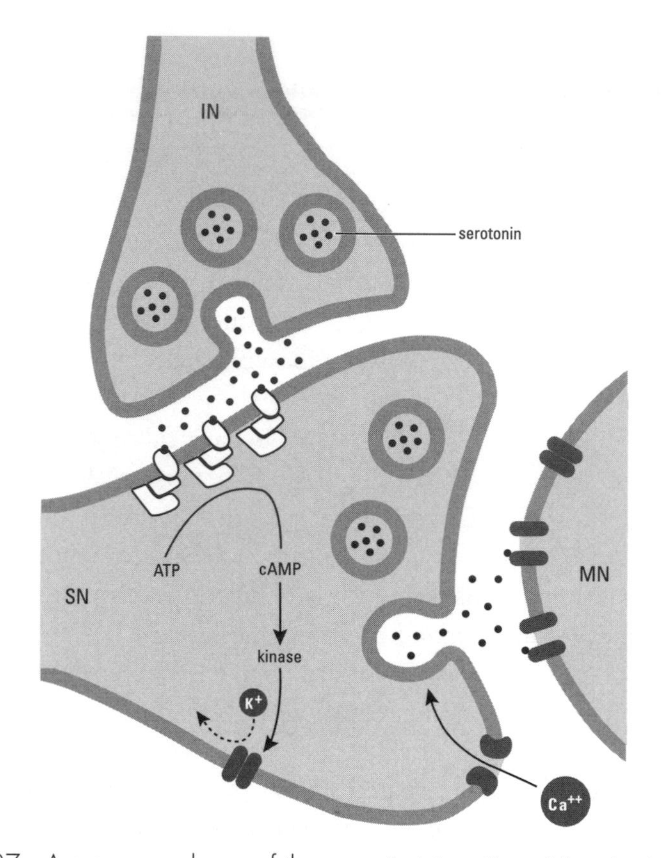

Figure 27 A summary scheme of the synaptic interaction giving rise to sensitization. Serotonin released from the interneuron terminal (IN) activates receptors in the sensory neuron terminal linked through a G-protein to adenylate cyclase, an enzyme that changes ATP to cyclic AMP. The kinase activated by cyclic AMP phosphorylates K^+ channels in the sensory neuron terminal membrane, thereby closing them. This results in prolongation of terminal activation and increased release of neurotransmitter from the sensory neuron terminal because more vesicles can bind to the membrane. With more neurotransmitter released, the motor neuron response is larger and gill withdrawal more extensive.

serotonin, which acts on sensory neuron terminals in classic neuromodulatory fashion: released serotonin activates receptors in the terminal membrane that are linked by a G-protein to the enzyme adenylate cyclase (see Chapter 3). This causes activation of the enzyme and the resultant second messenger, cyclic AMP, activates protein kinase A, which phosphorylates ion (K^+) channels in the terminal membrane.

Phosphorylation of this channel keeps the membrane potential of the terminal more positive (depolarized) for a longer time after an action potential comes down the sensory neuron axon. This happens because phosphorylation of the K^+ channels lowers their ability to allow K^+ ions to cross the membrane. Thus, repolarization of the membrane, caused by K^+ leaving the terminal, is slowed down after an action potential comes down the axon. With a prolonged depolarization, more Ca^{2+} can enter the terminal and promote neurotransmitter release. Here reflex alteration can be understood down to molecular and ionic levels!

What I've described so far relates to short-term habituation and sensitization. Both can be accounted for by alterations in transmitter release from the sensory neuron terminals. What about long-term habituation and sensitization? Are other factors involved here?

Anatomical studies were key to learning about these long-term changes. The sensory neurons in the ganglion were injected with a substance that filled the cell and allowed investigators to visualize sensory neuron axons and their terminals in the electron microscope. The big discovery was that sensory terminals vary between control, long-term-habituated, and long-term-sensitized animals. Terminals were fewer in long-term-habituated animals and more in long-term-sensitized animals; hence, fewer synapses exist between sensory and motor neurons in long-term habituation and more in long-term sensitization.

Alterations in individual synapses were noted as well, including changes in their size and the number of vesicles per synapse. These observations indicate that long-term habituation and sensitization bring about structural changes in neurons and synapses. This implies that new protein synthesis or degradation happens in the neurons because of these processes, and even that genes are turned on or off. Details are sketchy, but the evidence is compelling that kinases within nerve and other cells can alter protein synthesis and degradation, and gene activation and inactivation.

The experiments on *Aplysia* thus provide experimental support for the older ideas that short-term memories reflect alterations in ongoing neural processes and that long-term memories reflect alterations in neural structures. The *Aplysia* model explains elegantly what such changes might entail and how they can come about. In us, a region of the brain called the *hippocampus* is essential for consolidating short-term memories. Destruction of the hippocampi (we have two, one on each side of the brain) results in individuals who cannot remember things for more than ten or fifteen minutes (see Chapter 8). From the many studies of hippocampal mechanisms investigators are finding that similar mechanisms operate in the hippocampus as in the abdominal ganglia of *Aplysia*, when stimuli to the hippocampus result in long-term physiological changes.

The foregoing might suggest to you that only invertebrates have given us powerful insights into neural mechanisms. This is far from the truth. Indeed, we have learned much from studying vertebrates, and neuroscientists now routinely investigate neural mechanisms in higher mammals such as cats and monkeys. Again, though, neuroscientists have often turned to simpler vertebrates, such as fish or frogs, to investigate neuronal mechanisms or well-defined and perhaps simpler parts of the nervous system. My own research is on the vertebrate retina, a part of the brain displaced into the eye during development of the embryo. Its organization and anatomy are well understood. Furthermore, certain animals such as the mudpuppy, a salamander-like animal, have large cells for facilitating recordings from the neurons. Other cold-blooded vertebrates, like fish, have extraordinarily large retinal neurons of one type, and much has been gleaned from these neurons. The point, then, is that neuroscientists take advantage of the enormous variety of animals that have evolved on this planet in their efforts to understand the brain. Neurobiological mechanisms are similar in all species, which makes this approach useful. Ultimately, though, it is detailed knowledge of the human brain that we seek.

CHAPTER 5

Brain Architecture

Christina was a strapping young woman of twenty-seven, given to hockey and riding, self-assured, robust, in body and mind. She had two young children, and worked as a computer programmer at home. . . . She had an active, full life—had scarcely known a day's illness. Somewhat to her surprise, after an attack of abdominal pain, she was found to have gallstones, and removal of the gallbladder was advised.

She was admitted to hospital three days before the operation date, and placed on antibiotics for microbial prophylaxis. This was purely routine, a precaution, no complications of any sort being expected at all. Christina understood this, and being a sensible soul had no great anxieties.

The day before surgery Christina, not usually given to fancies or dreams, had a disturbing dream of peculiar intensity. She was swaying wildly, in her dreams, very unsteady on her feet, could hardly feel the ground beneath her, could hardly feel anything in her hands, found them flailing to and fro, kept dropping whatever she picked up.

She was distressed by this dream—("I never had one like it," she said. "I can't get it out of my mind.")—so distressed that we requested an opinion from the psychiatrist. "Pre-operative anxiety," he said. "Quite natural, we see it all the time."

But later that day *the dream came true*. Christina did find herself very unsteady on her feet, with awkward flailing movements, and dropping things from her hands.

The psychiatrist was again called—he seemed vexed at the call, but also, momentarily, uncertain and bewildered. "Anxiety hysteria," he now snapped, in a dismissive tone. "Typical conversion symptoms—you see them all the while."

> But the day of surgery Christina was still worse. Standing was impossible—unless she looked down at her feet. She could hold nothing in her hands, and they "wandered"—unless she kept an eye on them. When she reached out for something, or tried to feed herself, her hands would miss, or overshoot wildly, as if some essential control or coordination was gone.
>
> She could scarcely even sit up—her body "gave way." Her face was oddly expressionless and slack, her jaw fell open, even her vocal posture was gone.
>
> "Something awful's happened," she mouthed, in a ghostly flat voice. "I can't feel my body. I feel weird—disembodied."
>
> —Excerpted from Oliver Sacks, *The Man Who Mistook His Wife for a Hat*

Christina had lost all position sense—sensory information coming from the muscles, joints, and tendons telling the brain the status of her limbs and trunk, including their positions. Such information is called proprioceptive and is largely unconscious. We are unaware of most of this sensory input from the limbs and body, but it is essential for the brain to "sense" the body. As Christina noted, "I lose my arms. I think they are in one place and I find they're [in] another."

A quite peculiar and selective inflammation of the proprioceptive axons had occurred all along Christina's spinal cord and brain. Over the next few days the inflammation gradually subsided, but the axons did not recover. Christina had permanently lost virtually all proprioception. However, Christina did gradually manage to deal with her deficit, by using her eyes and to some extent her ears to monitor her movements, to keep track of her limbs, and to modulate her speech. But it was an excruciatingly slow process. It took nearly a year before she could leave the hospital, rejoin her family, and resume her computer work. She still was by no means normal but had learned to compensate. How did she feel? It was difficult for to describe her state except in terms of her other senses. "I feel my body is blind and deaf to itself . . . it has no sense of itself."

In vertebrates, the nervous system is divided into two parts: the central nervous system, made up of the spinal cord and the brain proper; and the peripheral nervous system, consisting of all the nerves and nerve cells that lie outside of the spinal cord and brain. The peripheral nervous system is largely made up of motor axons and sensory neurons. The axons of the motor neu-

rons carry the instructions from the central nervous system out to the muscles of the head, face, limbs, and body, allowing us to make purposeful movements and to act. Sensory neurons, carrying information into the central nervous system from the various specialized sense organs, inform the brain about the rest of the body and the world beyond. We are not aware of all of the sensory information coming into the brain, including most proprioceptive input. Without proprioceptive sensory input, the brain is isolated from much of what is going on in the body, as happened to Christina. If other sensory systems are lost, we would be isolated completely from the outside world. Helen Keller was blind and deaf from an early age. Through the heroic efforts of her teacher, Anne Sullivan, contact with her was made through her sense of touch. Without the sense of touch, Helen Keller would have always been isolated from other people.

Although neuronal mechanisms in the central and peripheral nervous systems are virtually identical, there is a surprising difference in the glial cells of these systems in mammals that has important consequences for us. In the peripheral nervous system, the glial cells that surround nerve cell axons and form the myelin sheath are called Schwann cells, named after their discoverer, Theodore Schwann. He is best known for his theory—announced in 1839—that all animals are formed of cells. In the central nervous system, the glial cells that form myelin are known as oligodendrocytes. Why the difference in glial cells is important relates to the regrowth of axons after they are cut or injured. Whereas axons in the peripheral nervous system will regenerate, axons in the central nervous system of mammals will not, a difference related to the surrounding glial cells.

If a limb or other part of the body served by the peripheral nervous system is severed and then surgically reattached, function and sensation can make a remarkable recovery. This is a slow process requiring many months, but sensory and motor neuron axons eventually regenerate and make appropriate connections. Axons in the central nervous system, in contrast, do not regrow after injury or cutting. This is best exemplified by spinal cord injuries that cut or crush nerve cell axons. Paralysis and loss of sensation occur below the level of injury, and this condition is permanent. Such individuals never can walk again, and if the spinal cord injury is just below the neck, they may also lose the use of their arms. They may also need assistance breathing and require a respirator. The actor Christopher Reeves, whose spinal cord was crushed in a riding accident, provides an example of this tragic injury.

Albert Aguayo and his colleagues in Montreal have demonstrated that if severed central axons are put in contact with Schwann cells, they will regenerate. The experimenters remove a piece of nerve from an animal's leg or another part of its peripheral nervous system, which prompts the axons in the nerve to degenerate. Then the investigators place the remaining Schwann cells around a severed nerve, such as the optic nerve, of the central nervous system. The result is that some axons in the optic nerve will regenerate and reinnervate part of the brain and some sight is partially restored. The importance of such findings for the possibility of treating spinal cord and brain injuries cannot be overstated.

Why do Schwann cells allow for axonal regeneration but oligodendrocytes do not? This is still not well understood. Two theories, not mutually exclusive, are being tested. One proposes that Schwann cells release one or more substances that promote axon regeneration; the other theory states that oligodendrocytes release a factor that forbids the regeneration of axons. Some evidence supports both theories, and both situations may hold. Also, central nervous system axons do regenerate in many cold-blooded vertebrates, so comparing oligodendrocytes from mammalian and nonmammalian species may provide more clues to the glial cell factors involved. The critical and surprising conclusion is that axonal regrowth is regulated by glial cells, not by the neurons themselves.

The Central Nervous System

Figure 28 is a drawing of the central nervous system—the spinal cord and brain—of a human viewed from the back side. The spinal cord is housed in the vertebral column; the brain is in the skull. Unlike most invertebrate nervous systems, which are distributed along the animal, the nervous system in vertebrates such as ourselves is highly centralized. In us, all higher nervous system functions, including movement initiation, perception, memory, learning, and consciousness, are carried out within the brain. The spinal cord, in contrast, serves three major roles: Simple reflexes, such as the knee-jerk reflex in which a tap just below the knee makes the lower part of the leg kick out, are mediated within the spinal cord; the neural circuitry for rhythmic move-

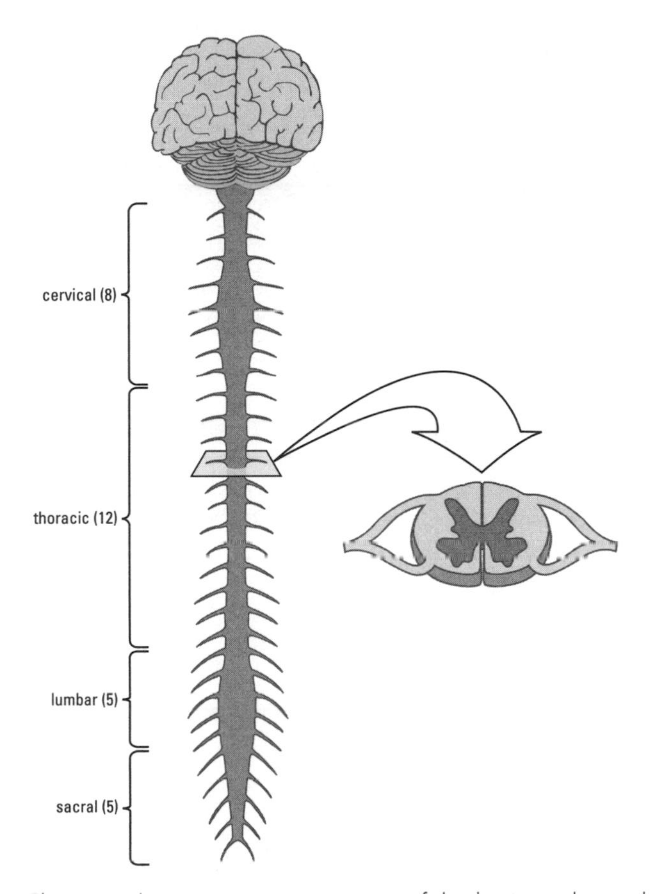

Figure 28 The central nervous system consists of the brain and spinal cord. Bundles of axons extend from the spinal cord in pairs—thirty pairs in all—and carry sensory information into the spinal cord, or motor information out from the spinal cord to the muscles. A cross section of the spinal cord in the thoracic region is indicated on the right. Four regions along the spinal cord are recognized—cervical, thoracic, lumbar, and sacral. The number of nerve bundles exiting from each region is indicated in parentheses.

ments such as walking or scratching are found in the spinal cord; and the cord itself serves as a conduit for sensory information from the periphery to the brain and for motor information from the brain to the motor neurons found in the spinal cord.

As shown in *Figure 28*, thirty pairs of nerve bundles extend from the spinal cord. These nerves consist of sensory and motor axons. Sensory information from the body comes into the spinal cord by way of axons in these nerves; motor information goes to the muscles via other axons in the nerves. The left member of a pair of nerves extending from the spinal cord serves a portion of the left side of the body, the right member of the pair serves a portion of the right side of the body.

Much of the sensory information entering the spinal cord is called *somatosensory*—conveying information from touch, temperature, pressure, and pain receptors in the skin and deep tissues of the limbs and trunk. In addition to the thirty pairs of spinal cord nerves, twelve pairs of nerves enter the brain directly. These cranial nerves carry sensory and motor information related to the head. Three cranial nerves carry only sensory information, visual, auditory, or olfactory; three are devoted to controlling eye movements; and four are a mix of sensory and motor axons innervating the face, tongue, neck, and jaw. Finally, two cranial nerves innervate internal organs such as the heart, lungs, and digestive system.

Spinal Cord

A slice through the spinal cord is depicted in *Figure 29*. In the cord's interior is a butterfly-shaped region that stands out from surrounding regions by its grayish hue (termed *gray matter*). The cell bodies of neurons and their dendrites are in this interior region, and most of the synapses are made here also. The surrounding, whiter region (*white matter*) contains bundles of axons running up and down the cord. The axonal bundles are whitish because of the myelin sheaths covering them.

Sensory information comes into the cord dorsally—from the back side—while motor information exits the cord ventrally—from the front side. The cell bodies of sensory neurons are collected together in ganglia that lie outside the cord. The sensory neuron axons entering the spinal cord can do any one of three things. The can ascend the cord, carrying sensory information to higher cord levels or to the brain itself; they can synapse on interneurons found within the gray matter; or they can synapse directly on motor neurons found ventrally in the gray matter.

When a sensory neuron axon directly innervates a motor neuron, as in *Figure 29*, a simple reflex circuit is created. The sensory neuron directly activates a motor neuron, and the result is a behavior independent of the rest of the nervous system. The knee-jerk reflex is such a reflex in us; a one-synapse reflex is called a *monosynaptic reflex*. Reflexes may involve interneurons and hence be polysynaptic. Reflex circuitry can become quite complicated, involving many interneurons and several levels within the spinal cord.

The axons that run up and down the cord are segregated according to whether they are sensory or motor axons and what type of sensory

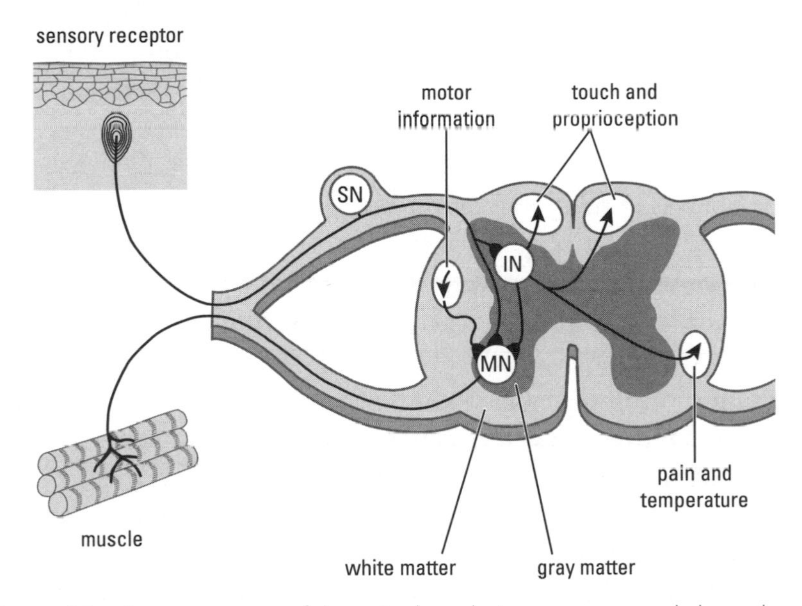

Figure 29 A cross section of the spinal cord. Axons run up and down the cord around the periphery of the cord in the white matter. The neurons, dendrites, and synapses are in the central gray matter. Sensory information enters the cord dorsally, and synapses of the sensory neurons (SN) may be made onto interneurons (IN) or directly onto motor neurons (MN). The latter situation is the basis of a simple monosynaptic reflex. Motor neuron axons exit the cord ventrally to innvervate muscles. Sensory information travels up the cord mainly via axons of the interneurons. These axons run along specific paths in the white matter depending on the type of sensory information they are carrying. Interneurons may innervate motor neurons, and the motor neurons also receive input from axons descending down the cord (left side of drawing). For simplification, information entering and exiting the cord is shown only for one side, and the interneuron is shown having multiple axons.

information they are carrying. So axons carrying pain and temperature information are in one region of the cord, touch in another region, and so forth. Motor axons are also segregated, descending in areas separate from sensory areas, some of which are indicated in *Figure 29*.

A curious feature of the organization of the vertebrate nervous system is that much of the sensory information ascending the spinal cord does so on the opposite side of the cord from where it enters, and it eventually reaches the brain on this opposite side. Hence information from the body's left side is processed mainly on the brain's right side, and vice versa. The same holds for motor information. The brain's right side initiates movements for the body's left side, and vice versa. Why one side of the brain controls the opposite side of the body is a mystery, and no one has a compelling explanation of why this is so. Its consequences are that an injury on one side of the brain or spinal cord affects sensation or motor control on the body's opposite side: A stroke on the brain's left side paralyzes the body's right side, and vice versa.

The sensory information ascending the spinal cord described so far is all conscious information; we are aware of it. It reaches brain regions that participate in conscious perception. The cord also has ascending sensory information of which we are not aware. This sensory information from our muscles, joints, and tendons is critical for efficient and coordinated movement and essential for proper brain function. It tells the brain the position of our limbs and their status. The degree of muscle contraction is signaled to the brain, for example, but we are not aware of such sensory information; it does not enter our consciousness. This information is called *proprioceptive* after the Latin *proprius*, meaning "one's own." Without it, patients are severely debilitated, as the story of Christina at the beginning of this chapter illustrates.

Proprioceptive information runs up the spinal cord dorsally. In the late stages of syphilis these tracts partially degenerate, and patients suffering from this degeneration typically have exaggerated movements like high-stepping when they walk. Fortunately, with the advent of antibiotics only rare cases of this degeneration are seen today. But even in the 1950s, when I was a medical student, patients with the characteristic gait of this degeneration were not uncommon.

Another degeneration involving the spinal cord that is no longer a problem, at least in the developed world, is poliomyelitis. The polio virus invades motor neurons in the ventral regions of the spinal cord and destroys them, leading to the paralysis of the muscles these neurons innervate. All muscles of the limbs and trunk are controlled by motor neurons in the spinal cord. They are, in the words of Charles Sherrington, an eminent English scientist working early in this century who was particularly interested in spinal cord reflexes, the "final common pathway" regulating movement. They are acted on by axons descending the spinal cord carrying information from the brain, directly from sensory neurons entering the spinal cord, and by spinal cord interneurons. Their output determines the movement of a part of the body, and their loss results in paralysis of that part.

Finally, within the spinal cord is the circuitry underlying several rhythmic movements such as walking and scratching. The evidence for this comes from experiments on cats, which received a lesion in the lower part of the brain that separates this region and the spinal cord from higher brain regions. The animals can stand, and if placed on a treadmill will walk if a specific group of neurons in their lower brain stem is stimulated. These neurons appear to provide a signal, a command, to interneurons in the spinal cord, which then interact in such a way that they give motor neurons appropriate synaptic input for coordinated walking. We only vaguely understand the circuits underlying these rhythmic motor behaviors, but many investigators are trying to work out their circuitry.

The Brain

Figure 30 shows a longitudinal section through the human brain and part of the spinal cord. Essentially this section is a cut down through the middle of the brain, and it reveals several of the major brain structures. As expected, the structure of the brain is complex, consisting of a number of subdivisions that differ strikingly in their anatomy. And such a longitudinal section fails to reveal certain important brain regions.

To sort out the various structures, it is helpful to divide the brain into three regions: *hindbrain*, *midbrain*, and *forebrain*. The hindbrain emerges from the spinal cord and has three main structures, the *medulla*, *pons*, and *cerebellum*. The midbrain sits between the hindbrain and forebrain, and in humans it is relatively small. The forebrain is by far the most prominent part of the human brain and is divided into two regions. One includes the *thalamus* and *hypothalamus* and the other includes the *basal ganglia* (not seen in *Figure 30*) and the *cerebral cortex*. The basal ganglia and cerebral cortex are called collectively the *cerebrum*. A prominent band of axons, the *corpus callosum*, lies centrally under the cerebral cortex and carries information between the brain's right and left sides.

The brains of cold-blooded vertebrates such as fishes or frogs provide insights into the evolution of the vertebrate brain (*Figure 31*). The brains of these animals have a small cerebrum relative to the rest of the brain. In most cold-blooded vertebrates the cerebrum is concerned primarily with the analysis of one sensory modality, smell. In mam-

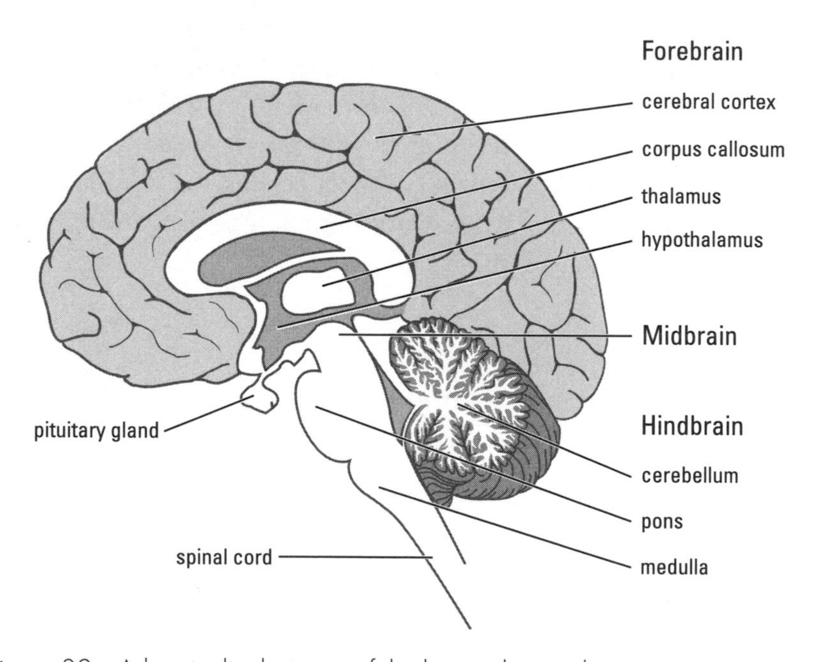

Figure 30 A longitudinal section of the human brain, showing its major structures. The hindbrain emerges from the spinal cord and consists of the cerebellum, pons, and medulla. The midbrain encompasses a relatvely small part of the primate brain, whereas the forebrain is very prominent. Major structures of the forebrain include the cerebral cortex, corpus callosum, thalamus, and hypothalamus. The pituitary gland is found just below the hypothalamus.

mals the cerebrum is greatly expanded and most higher neural functions are centered in the cerebral cortex. The larger forebrain, especially the cerebral cortex, is by far the most significant difference we observe when comparing higher mammals with cold-blooded vertebrates. The evolution of the vertebrate brain is thus mainly about the forebrain; it has evolved from a structure concerned mainly with analyzing just one sense to, in humans, the seat of sensation, memory, intelligence, and consciousness.

In cold-blooded animals, the midbrain is a much more prominent part, and the brains of many of these animals are dominated by a structure called the *tectum*. The tectum receives visual and other sensory input, and tectal neurons project to the spinal cord, where they synapse onto motor neurons. The tectum is the key structure in non-

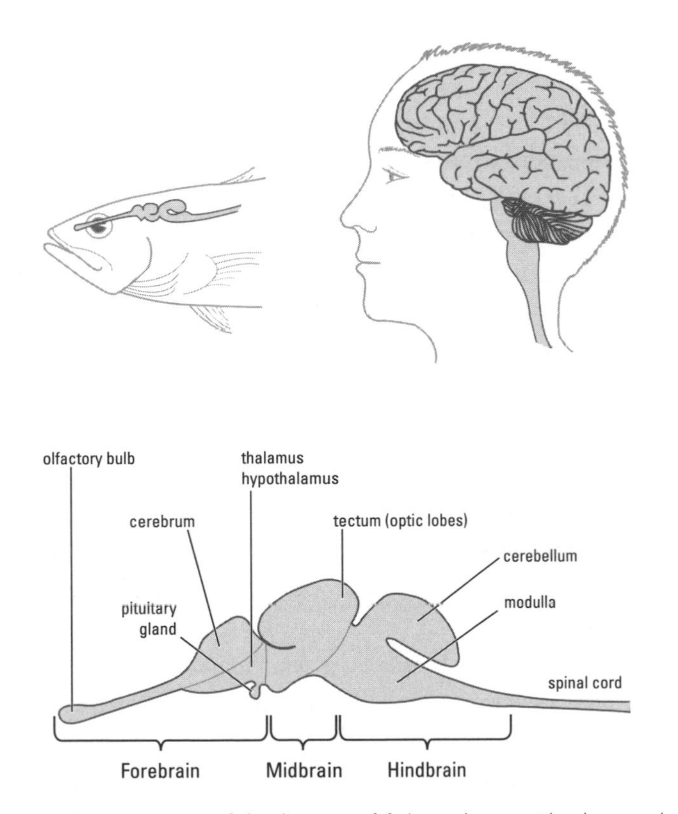

Figure 31 A comparison of the brains of fish and man. The human brain is very much larger, but the same structures exist in both brains (compare with Figure 30). However, the midbrain in the fish is relatively much larger and is dominated by the tectum.

mammalian species for integrating sensory inputs and for controlling motor outputs.

In higher mammals, the midbrain and tectum are less important brain structures. Sensory processing, sensory-motor integration, and the initiation of motor activity are in the cerebral cortex. The midbrain and tectum in the mammal mediate noncortical visual reflexes like pupillary and eye movements. In higher mammals the tectum also helps to coordinate head and eye movements.

Medulla and Pons

The medulla links the spinal cord to the brain. Thus, running through the medulla are numerous ascending (sensory) and descending (motor) axon bundles. Half of the cranial nerves enter the brain in the medulla. But within the medulla are critical nuclei (clusters of neurons) concerned with regulating vital body functions, including respiration, heart rate, and gastrointestinal function. The bullet that killed President Kennedy entered this part of his brain and destroyed many of these vital regulatory centers. Without these critical nuclei, death inevitably ensues. Nuclei that control the head, face, eyes, and tongue are also in the medulla.

In addition to the discrete nuclei in the medulla, small groups of neurons are diffused throughout the medulla. These neurons make up what is called the *reticular formation*. They extend their axons widely throughout the brain, and their synaptic terminals typically contain neuromodulatory substances—monoamines or peptides. Reticular formation neurons exert their effects on virtually all parts of the brain. They are involved in arousal and controlling levels of consciousness; lesions in the reticular formation can cause animals, including humans, to lose consciousness or to fall into a stupor from which they cannot be aroused.

The pons (from the Latin word for "bridge") contains neurons that receive input from the cerebral cortex. They relay this information to the opposite side of the cerebellum. The pons thus serves as a switchboard between the cerebral cortex and cerebellum and mediates the crossing of most motor information from one side of the brain to the other. Some reticular formation neurons are also found in the pons.

Hypothalamus

The hypothalamus and the medulla are the brain's principal regulatory centers. Nuclei in the hypothalamus mostly regulate basic drives and acts such as eating, drinking, body temperature, and sexual activity. The hypothalamus also plays a role in emotional behavior, which will be discussed in detail in Chapter 9. Lesions in the hypothalamus, or stimulation of parts of it, can lead to irritability or even aggressive behavior. Yet lesions or stimulation of other parts of the hypothalamus can lead to placidity.

The pituitary gland sits underneath the hypothalamus, and so another task for the hypothalamus is to regulate the pituitary gland's release of hormones. Hypothalamic neurons do this by releasing small peptides that promote or inhibit the release of the pituitary hormones. The pituitary hormones, by way of the bloodstream, regulate hormone release from glands found elsewhere in the body, such as the thyroid and adrenal glands; or they may exert direct effects on tissues. Examples of the latter are growth hormone and oxytocin, the mammary gland milk-releasing hormone. Hypothalamic nuclei, along with nuclei in the medulla, also help to control the autonomic nervous system. This system regulates internal organs, including the heart, digestive system, lungs, bladder, blood vessels, certain glands, and the pupil of the eye; it also will be discussed in detail in Chapter 9.

Cerebellum

The cerebellum coordinates and integrates motor activity. The command for a skilled movement comes from the cerebral cortex, but the cerebellum must coordinate a motor command with sensory information to ensure smooth movement. Thus, the cerebellum receives input from the cortex via the pons and sensory input from the spinal cord and other sensory systems. Much of the proprioceptive sensory information coming up the spinal cord goes to the cerebellum. The cerebellum compares the various inputs, integrates them, and sends signals to the motor neurons in the spinal cord to accomplish a smooth, coordinated movement.

Lesions of the cerebellum typically result in jerky, uncoordinated movements. Movement initiation may be delayed, or movements may be exaggerated or inadequate. The cerebellum also is active in the learning of motor tasks, such as riding a bicycle. Thus, cerebellar lesions can also interfere with learning new motor skills.

Thalamus

The thalamus has numerous nuclei whose role is to relay sensory information to the cerebral cortex. Other thalamic nuclei send information to the cortex about motor activity. Individual nuclei thus convey specific sensory or motor information. The lateral geniculate nucleus, for example, transmits visual information to the cortex. The optic nerve terminates in this nucleus, and the neurons in the lateral geniculate nucleus then project to the area of the cortex that processes visual information.

The thalamic nuclei receive input back from the cortex and also from the reticular formation in the medulla. These nonsensory inputs control the flow of information from the thalamus to the cortex. Thus, a key role carried out by the thalamus is to gate information flow between the spinal cord and lower brain structures and the cerebral cortex.

Basal Ganglia

Lateral to the thalami on both sides of the brain are five prominent nuclei known as the *basal ganglia*. These are concerned primarily with the initiation and execution of movements. The position of the basal

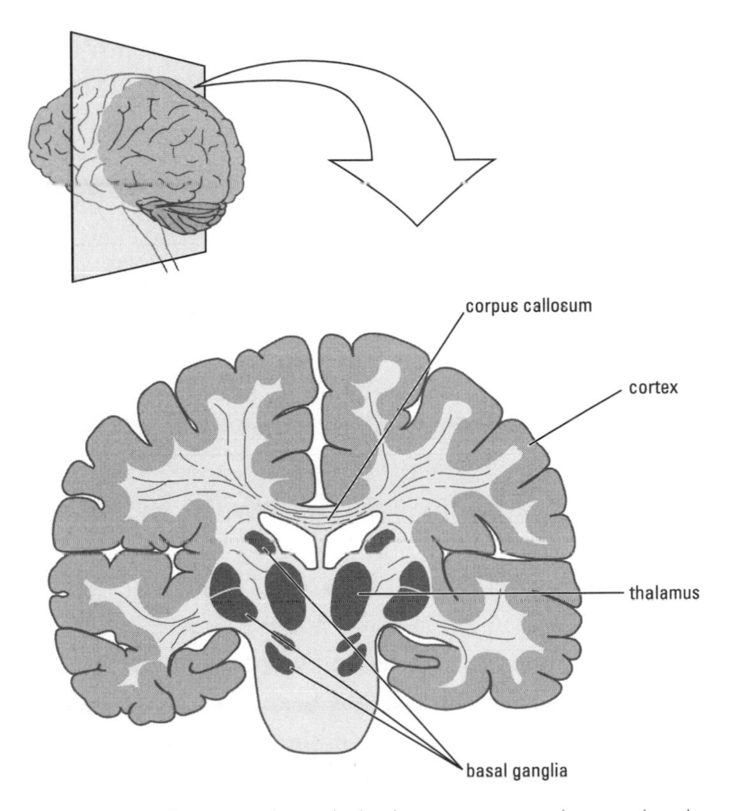

Figure 32 A vertical section through the brain cut according to the diagram shown above. The cortex is a dense cellular layer, about 2 mm thick, covering the two cortical hemispheres. The infoldings of the cortex serve to increase the brain's surface area and, therefore, the amount of cortex. The corpus callosum is a thick band of axons that connects the two hemispheres. The basal ganglia and thalamus are subcortical structures.

ganglia is shown in *Figure 32*, which represents a vertical section through the middle of the brain. The basal ganglia receive input from the cortex and feed information back to the cortex by way of the thalamus. Lesions in the basal ganglia cause characteristic movement abnormalities. Tremors and repetitive movements are frequently seen in such patients, and individuals with basal ganglia defects have difficulty in initiating movements. Rigidity of the limbs is another common symptom.

One disease of the basal ganglia is, as we have noted, Parkinson's disease, and another is Huntington's disease, which is inherited and is also related to the loss of specific neuroactive substances within certain ganglia. In Parkinson's disease dopaminergic neurons are lost, and in Huntington's disease GABA and acetylcholine are lost; neurons containing these substances characteristically degenerate. Early in Huntington's disease, which typically begins in middle age, patients experience small but uncontrolled movements of the arms, legs, torso, and face. These spontaneous movements gradually worsen, the patients have difficulty swallowing, and they lose their balance and become very unsteady until they are confined to bed. Accompanying the movement disorder are mood swings, depression, irritibility, and eventually loss of cognition and a dementia. Death generally ensues fifteen to twenty years after onset of the disease.

Approximately 25,000 Americans are affected with Huntington's disease, with another 125,000 individuals at risk for it. The disease is inherited in a dominant fashion; that is, one copy of the defective gene from either the father or mother will cause the disease. Thus, 50 percent of the offspring of an affected individual will inherit the disease. The gene that causes the disease has been isolated and its sequence determined. From the gene sequence, the protein structure for which it codes can be predicted; but the predicted protein is unlike any known protein, hence its function is a mystery. The protein is widely scattered throughout the nervous system, which suggests that it plays a significant role, but what that role may be is presently unknown.

Cerebral Cortex

Without doubt, the most important part of the brain for us is the cerebral cortex. It is in the cortex that virtually all of our higher mental functions are localized. The initiation of skilled movements, perception, consciousness, memory, and intelligence all depend critically on the cerebral cortex. The cortex is divided into two hemispheres, each of which is subdivided into four lobes termed *frontal*, *parietal*, *occipital*, and *temporal*. *Figure 33* shows a surface view of the left hemisphere of the human cerebral cortex.

The cortex's neurons are close to the surface, where they form a layer about 2 mm (less than 0.1 inch) thick that covers each hemisphere. To increase surface area, and thereby have more cortical neurons, the cortex is extensively infolded in higher mammals and ourselves. How this works is shown in *Figure 32*. In humans, the total area of the cortex is about 1.5 square feet if spread out, and under each square millimeter of cortex are about 100,000 neurons. Thus, the human cortex as a whole has about 10^{10} or 10 billion neurons.

Specific roles are carried out within each cortical lobe. The frontal lobes, for example, are concerned primarily with movement and smell, the parietal lobes with somatosensory information processing, the occipital lobes with vision, and the temporal lobes with hearing and memory consolidation. But each lobe consists of many subareas, whose roles vary widely. For example, within each of the lobes are areas devoted to the initial processing of one sensory modality or to the initiation of skilled movements. These primary sensory areas and the primary motor projection area are shown in *Figure 33*. A deep infolding,

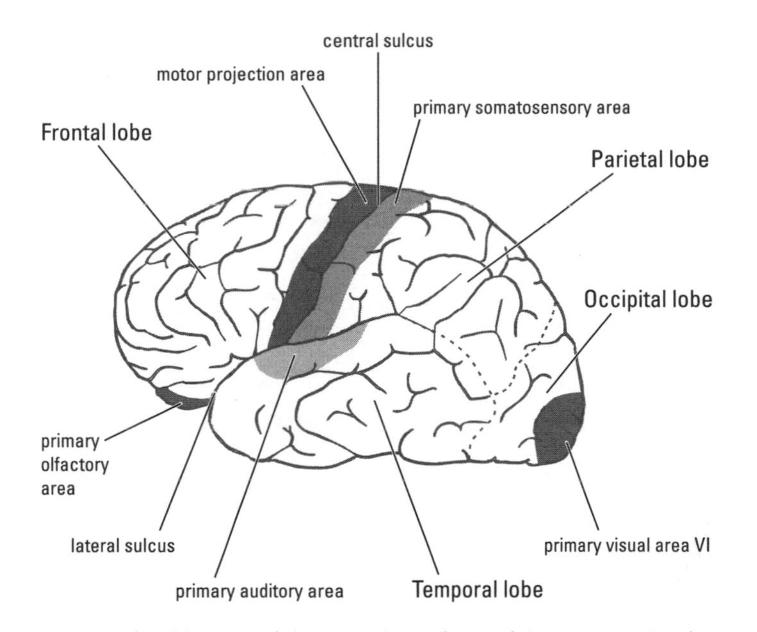

Figure 33 A left-side view of the outside surface of the cortex. The four cortical lobes—frontal, parietal, occipital, and temporal—are shown, as are the primary sensory-processing areas and motor projection area. Two of the deep infoldings (sulci) are indicated—the central sulcus separates the frontal and parietal lobes, and the lateral sulcus separates the frontal and temporal lobes.

called the central sulcus, separates the motor projection area from the primary somatosensory area, and also the frontal lobe from the parietal lobe. Another deep infolding, called the lateral sulcus, separates the temporal lobe from the frontal and parietal lobes. Many more infoldings are present in the human cortex than are shown in *Figure 33*.

Along the primary somatosensory cortex, sensory information from parts of the body is received and analyzed. Thus a rough representation of the body surface exists on the primary somatosensory cortex (*Figure 34*). This representation is, however, not proportional. More cortical area is devoted to body parts where sensation is more acute, such as the face and hand, and less cortical area is devoted to

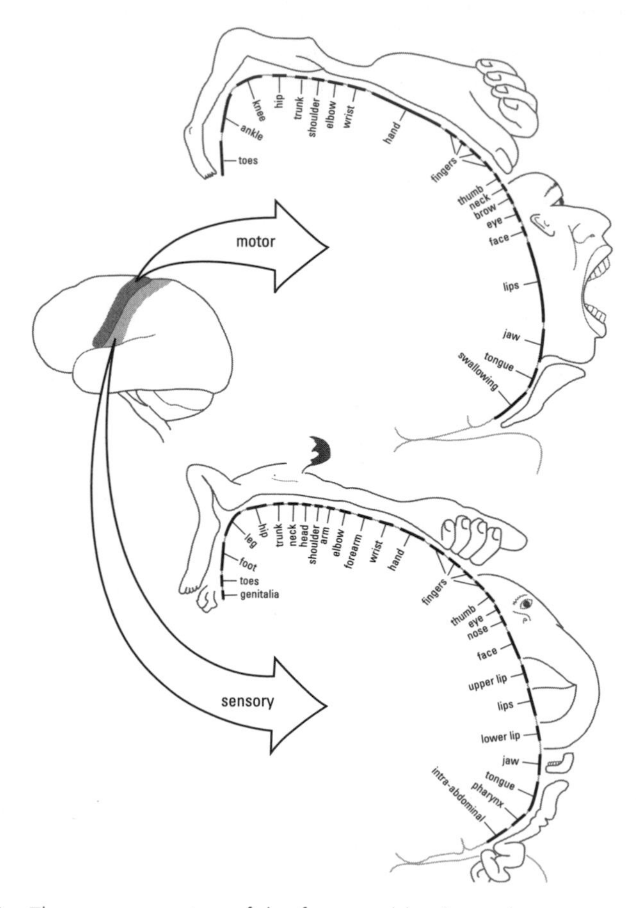

Figure 34 The representation of the face and body on the primary motor cortex (above) and primary somatosensory cortex (below). Areas such as the hands and face with more innervation—where sensation is more acute or that have finer movements—have larger cortical representations.

regions that receive less sensory innervation, such as the back or upper leg. The same situation holds for the primary motor projection area and the primary visual area. The hand and face occupy a disproportionate amount of the primary motor projection area, and the central (foveal) region of the retina, which mediates high-acuity vision, encompasses a large proportion of the primary visual area. One can relate the sensitivity of sensation or the degree of motor control to the amount of cortex it occupies or, simply, the number of neurons involved.

Away from the primary sensory or motor projection areas are secondary or association areas that participate in higher levels of processing or the processing of one aspect of a sensory stimulus. As many as thirty areas may be involved in the processing of visual information. Damage to the primary visual area, area V1, causes the loss of all visual perception. That is, a patient is not aware of seeing anything in that part of the visual field which projects to the damaged cortical area. The patient may exhibit visual reflexes that are mediated by lower brain centers. For example, if an object is brought rapidly toward the individual, he might blink or even move his head in an evasive way. But the subject will not be able to say why he blinked or moved his head or even that he saw something.

A lesion in an association area (i.e., V4) might lead to the loss of color vision, but leave other aspects of vision normal. The patient would see objects perfectly well, moving stimuli and so forth, but everything would be colorless. A lesion still farther away from area V1 might lead to the patient's inability to recognize someone. In the next chapter, I explore in depth how we see (which is my own area of research) and examine ideas on visual perception and the roles of area V1 and the visual association areas.

CHAPTER 6

Vision: Window to the Mind

Jonathan I. was a successful sixty-five-year-old artist, who painted abstract and colorful paintings. One day, while driving in the city, he was hit by a small truck on the passenger side of his car. He seemed unhurt but developed a severe and persistent headache, and so he went home. He slept exceptionally deeply that night and the next morning could not remember the accident. Furthermore, he soon discovered he could not read—letters all appeared like "Greek" to him—and furthermore he noticed he was unable to distinguish colors.

His physician arranged for him to be tested at a local hospital, which diagnosed him as having a severe concussion. His inability to read lasted for five days and then disappeared. However, his loss of color vision persisted and turned out to be permanent. For Jonathan this was devastating. His studio, which was filled with brilliantly colored paintings, was now only gray to him. His own paintings appeared strange to him, and they had lost their meaning. The weeks that followed were extraordinarily depressing.

Neurologists surmised that Jonathan had suffered a small stroke in that part of the cortex concerned with color vision processing. The rest of his vision was quite unimpared. He could see objects as sharply as before, could make accurate judgments of gray scales, and he could see moving objects perfectly well. He could read and draw accurately, but everything was without color. He could not even imagine colors or visualize them in his mind!

Jonathan slowly began to paint again, and to do sculpture—but all in black and white. He also became more and more of a night person, feeling comfortable in the night world of muted colors. He arose earlier and earlier, to work and to enjoy the night, when he felt equal or

> even to some extent superior to normal people. Gradually his sense of loss left him; and he even came to feel "privileged"—that he saw a world heightened in form, without the distraction of color.
>
> —Adapted from Oliver Sacks, *An Anthropologist on Mars*

My father was an ophthalmologist, and my brother is one. One might surmise that my interest in vision is inherited. But the person who most influenced my choice of careers and instilled in me an avid interest in learning about visual mechanisms was George Wald, my mentor at Harvard.

Wald taught biochemistry to undergraduates at Harvard, and the experience of learning from him was an eye-opener, pun intended. I was excited to discover that we can understand biological processes at the molecular level. I remember to this day Wald's description of Albert Szent-Gyorgi's famous experiment with glycerinated muscle fibers. (Szent-Gyorgi was also the discoverer of vitamin C and won the Nobel Prize for this discovery.) Muscle fibers removed from an animal and soaked in glycerol could be induced to contract when adenosine triphosphate (ATP) was added to them. ATP is found in all cells and stores energy for cells, but that it could cause a dead muscle fiber to contract again seemed incredible to me.

Seniors at Harvard often carry out research in a professor's laboratory, so I approached Wald about working with him. He accepted me into his laboratory, and I began research in the summer between my junior and senior years. My project was to study vitamin A deficiency in rats and to evaluate the changes that take place in the photoreceptors when deprived of this vitamin. Wald had, in the mid-1930s, shown that vitamin A is a key component of light-sensitive molecules found in photoreceptors. Indeed, he received a Nobel Prize for his discovery of the role of vitamin A in vision. But exactly what happens to photoreceptors when deprived of vitamin A was still not well understood. This was the subject of my undergraduate project, which eventually turned into a Ph.D. thesis, although not directly. After graduating from Harvard, I began medical school, but the lure of research kept bringing me back to Cambridge and to Wald's laboratory. After two years of medical school, I took a leave of absence to spend a year in research. One thing led to another, including a Ph.D. degree, and more than thirty-five years later I am still on that leave of absence.

After I left Wald's laboratory, my interest in the visual system moved from the photoreceptors and how light is captured by these cells to how the second-order and third-order cells in the retina process signals from the photoreceptors (see Figure 35)*. That the retina is a part of the brain, pushed out into the eye during embryonic development, has long been known. Substantial visual processing occurs in the retina; that is, the visual message coming out of the eye by way of the optic nerve axons is complex. It communicates much*

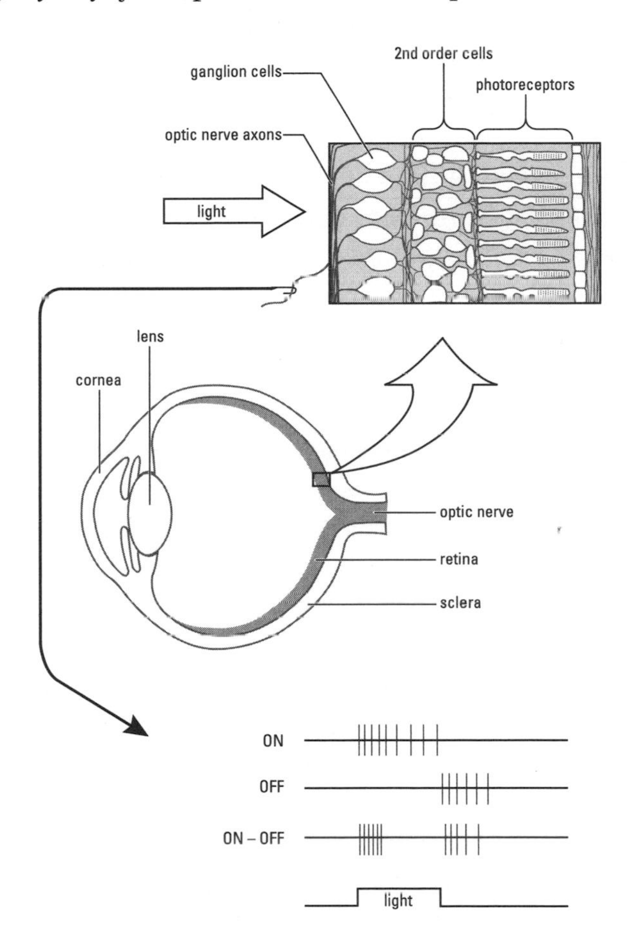

Figure 35 The retina, consisting of photoreceptors, second-order cells, and ganglion cells, lines the back (sclera) of the eye. The axons of the third-order ganglion cells run along the surface of the retina, forming the optic nerve where they exit the eye. Light enters the eye via the transparent cornea and is focused on the retina by the lens. Recordings from single optic nerve axons (bottom) show three basic types of responses when the retina is diffusely illuminated. Some axons respond when the light is on, others when the light is turned off, and some at both on-set and off-set of the light.

more than simply that a light is on. Indeed, since H. K. Hartline's experiments on the frog retina in the late 1930s, it had been known that some axons respond when light illuminating the retina goes on (they give an ON response), others after light goes off (OFF response), and others briefly at the onset and cessation of light (ON-OFF response).

By the early 1950s it was clear that most individual optic nerve axons respond to stimulation of only a small region of the retina, about 1 mm in diameter. This region is termed the receptive field, *and the fiber's response depends on where the light is shone in the receptive field. The fields are characteristically divided into two concentric and antagonistic zones. Stimulating the center of the doughnut produces an ON or OFF response, whereas stimulating the surrounding region elicits the opposite response. In addition, the center and surrounding regions are antagonistic—simultaneous stimulations of these regions cancel most of the response. In other words, the ganglion cells giving rise to the optic nerve axons respond less well to diffuse illumination; to activate them vigorously requires that illumination be positioned precisely on the retina.*

What did all of this mean and how was this processing accomplished? I spent the next twenty-five years of my life studying the question and am still vitally interested in the subject, although more recently we have been pursuing other questions, such as how the retina develops and how the cells of the retina talk to one another.

We began our studies of retinal processing by examining how the retina is wired. What do the cells look like and how do they connect with one another? Once we understood a bit about how information flows through the retina, we turned to recording the responses. This afforded insights into how cells code information and the different cells' roles. Combined with anatomical data, the physiological results permitted us to propose schemes for how the retina is organized. We could intelligently speculate on how ON and OFF responses are generated, how the center-surround receptive field organization of ganglion cells is established, and how complex visual information, like movement, is detected by retinal cells.

Next we began to ask how neurons talk to one another, what substances are involved, and how communication between cells can be modulated—not to mention the more recent question of how the retina develops, or what causes cells in the developing forebrain to form an eye. All our work has focused on the retina and the earliest stages of vision, but the act of seeing enlists many brain structures. From the eye, visual information passes through the middle of the brain into the cortex, where as many as thirty to forty different areas may participate in visual processing. Somchow it all comes together so we live in a coherent and visually rich world. But how? We have begun to glimpse underlying mechanisms, which is what this chapter is about—current ideas on photoreception, visual processing, and visual perception. It all has relevance for understanding brain mechanisms.

Catching Photons

Most eyes have two types of light-sensitive photoreceptor cells, called *rods* and *cones*. Rods mediate dim light vision, whereas cones function in brighter light and mediate color vision. Usually there is just one type of rod but several types of cones in vertebrate eyes. Humans, for example, have three types of cones—one that responds best to red-yellow light, another to green light, and the third to blue light. Rods and cones are elongated cells with a specialized outer segment region and an inner segment and synaptic terminal (*Figure 36*). The outer segment regions consist of numerous membrane infoldings, usually pinched off from the outer membrane in rods, but still connected to the outer membrane in cones.

Photoreceptors "see" because they contain light-sensitive molecules, known as *visual pigments*, in the cell's outer segment. Light is captured (absorbed) by these molecules, and this leads eventually to the excitation of the photoreceptor cell. The light-sensitive molecules are called *pigments* because they absorb certain wavelengths of visible light and hence have color. The visual pigment in rods, called *rhodopsin*, absorbs blue-green light best; it is most sensitive to blue-green light. It captures red and blue light less well; because it lets red and blue light escape, it thus appears purple to us. If one removes the retina from an eye of a dark-adapted animal that contains abundant rods (and most animals, including ourselves, have many more rods than cones), the entire retina has a reddish-purple hue. Indeed, the original name for the rod pigment was visual purple.

Light does two things when captured by the visual pigment

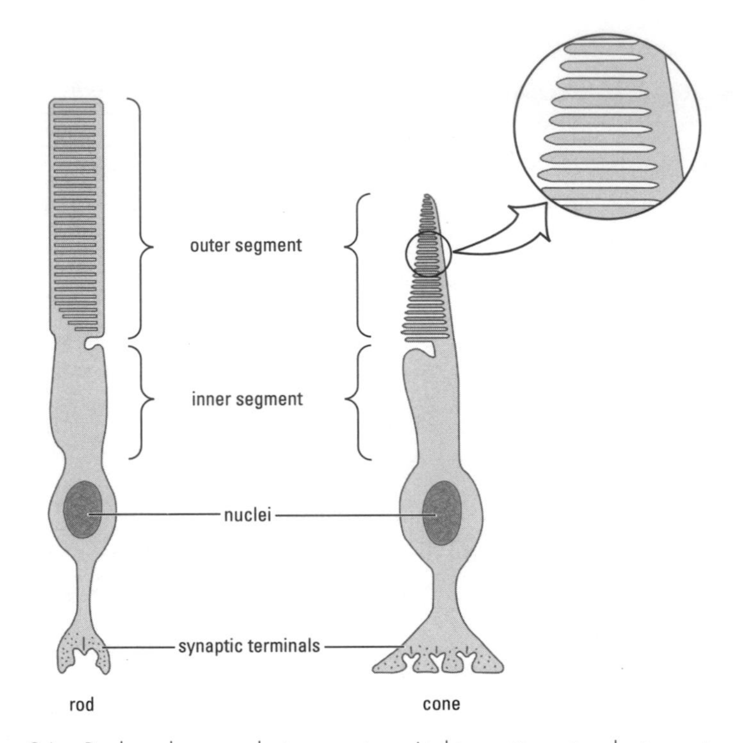

Figure 36 Rod and cone photoreceptors. Light-sensitive visual pigment molecules are present in the membraneous discs (rods) or infoldings (cones) of the outer segment regions of the cells. Information is passed on to second-order cells at the synaptic terminal regions of the cell.

molecule. First, it activates the molecule and excites the cell; second, it breaks down the molecule into component parts. All visual pigments consist of a large protein to which is bound a slightly modified form of vitamin A: vitamin A aldehyde or (its other name) *retinal*. When bound together, the molecule is sensitive to light in the visible range of the spectrum—between deep blue and far red light or between 400 and 700 nanometers. When split, the two components of the molecule absorb mainly ultraviolet light, which is invisible to us. The breakdown or bleaching of the visual pigment molecules in the light inactivates or desensitizes the photoreceptors. So when you go from bright light into a dark theater, it takes many minutes for your eyes to adjust. What you are waiting for is the resynthesis of the visual pigments in the photoreceptor cells—a process known as *dark adapta-*

tion—which restores the light sensitivity of the photoreceptors. After bright-light adaptation, cones dark-adapt in five or six minutes, but rods require up to thirty minutes to complete dark adaptation.

Individuals deficient in vitamin A are less sensitive to light than are well-nourished ones. This condition is known as *night blindness* because it is most obvious at night, and this is what I studied in vitamin A–deficient rats as a student with George Wald. Yet both rods and cones are affected in vitamin A deficiency—both are less sensitive to light and hence their visual pigments require vitamin A. What is different between rod and cone visual pigments and the three cone pigments is the proteins. Subtle differences in these proteins give the visual pigment molecules somewhat different properties, including their color sensitivity, i.e., the wavelength of light they best absorb.

Proteins are coded by genes, and so the genes coding for the rod and cone visual pigment proteins are different. Defects or alterations in these genes lead to significant visual abnormalities. Color-blind individuals either have lost a gene or have a defective gene for one or another of the cone visual pigment proteins. Red-blind individuals are missing or have an altered red-yellow visual pigment gene, green-blind individuals are missing or have an altered green-sensitive pigment gene, and blue-blind individuals are missing or have an altered blue pigment gene. The genes for the red- and green-sensitive pigments are on the X chromosome. Since males have just one X chromosome and females have two X chromosomes, red and green color blindness is much more common in males than in females. The reason for this is that even with only one good chromosome, color vision will be normal; the photoreceptor cell can still make a normal pigment. Since females have two X chromosomes, they can have one defective gene and one good gene and have normal color vision. Males have only one X chromosome; if that chromosome has a defective color pigment gene, the individual will be color-blind. Red-green color blindness is thus described as sex-linked.

It is important to note that most individuals we call color-blind still see colors. That is, if a person is unable to make one of the three cone visual pigments because of a defective gene (which is by far the most common situation) he or she still has two other cone types, i.e., green-and-blue-, red-and-blue-, or red-and-green-sensitive cones. With two cone types, color discriminations can be made, although such individuals cannot distinguish colors as well as a normal person with all three cone types. The color blindness exhibited by Jonathan I., described at the beginning of the chapter, was caused by a deficit in color vision processing in the brain, not by an alteration in his cones. He was totally color-blind; he could make no color discriminations at all.

Alterations in the gene for the rod pigment, rhodopsin, can lead to a disease called *retinitis pigmentosa*. People with this disease start off life with normal vision, but then the rods degenerate, first in the periphery of the retina and eventually throughout the retina. As the rods die, patients first lose the ability to see well in dim light, but eventually the cones die too, for reasons unknown, and all vision is lost. Why the rods die gradually, over the course of many years, is also not known. People with this genetic disorder usually begin to notice diminished visual sensitivity in their twenties or thirties. Complete blindness may come in the fifties or sixties.

Early Processing of Visual Information: The Retina

Lining the back of the eye is the retina, a thin layer of brain tissue that consists of four major types of neurons in addition to the photoreceptors. Within the retina two levels of processing happen: one between the photoreceptors and second-order cells, and the other between second- and third-order cells. The third-order cells (termed ganglion cells) are the retina's output cells; all visual information passes from the eye to the rest of the brain by way of the ganglion cell axons that make up the optic nerve.

The nature of retinal processing can be deduced by recording from the optic nerve axons—which means listening to the message being sent from the eye to the rest of the brain (*see Figure 35*). Two kinds of messages are being transmitted: One reflects outer retinal processing, the other inner retinal processing. Let's first consider outer retinal processing and the activity of the two ganglion cell subtypes that send the information. Half are ON-center cells that respond vigorously when their receptive field center is illuminated, and the others are OFF-center cells that respond vigorously when light is turned off in the receptive field center. The responses of such cells are shown in *Figure 37*. Thus, two pathways are established in the outer retina; one carries ON information and the other OFF information. What might this mean?

Experiments with animals have shown that when the ON-center cells are incapacitated, the animals can no longer tell if a spot of light is intensifying or brighter than the background, but they do know if it

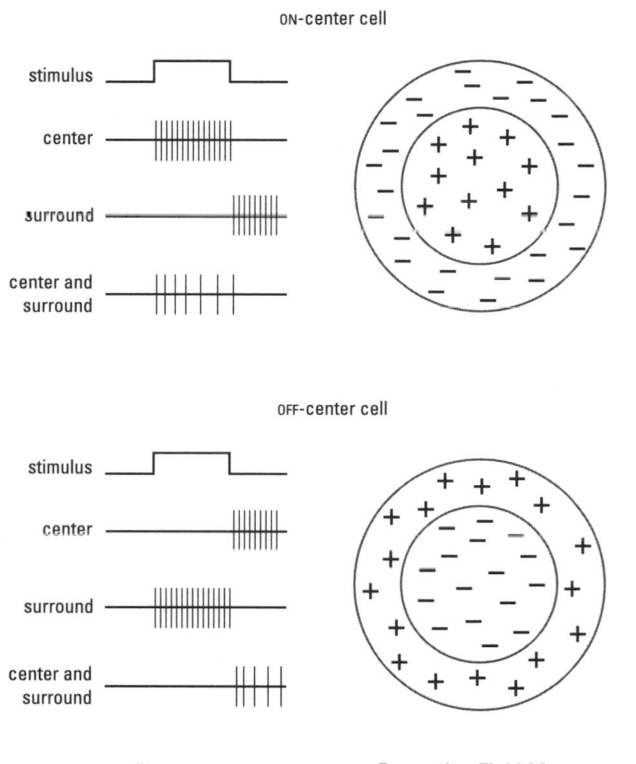

Figure 37 Responses (left) and receptive field maps (right) of an ON-center ganglion cell (above) and OFF-center ganglion cell (below). The response of the cell depends on which part of the receptive field is illuminated. For an ON-center cell, illumination of the receptive field center results in an ON response (+ symbol on the receptive field map); illumination of the surrounding region results in an OFF response (– symbol on the receptive field map). When both center and surround are illuminated, a weak center-like response is evoked in the cell. The OFF-center cell behaves similarly but in a mirror-image fashion. Central illumination gives rise to an OFF response; surround illumination to an ON response; and joint center and surround illumination to a weak OFF response.

is weakening or dimmer than the background. These data suggest that ON-center cells tell the rest of the brain about brighter illumination, whereas OFF-center cells signal diminished illumination. Thus, whether a light spot is brighter or dimmer than the background is signaled by different pathways. From these and a slew of other experiments comes a recurrent theme with regard to visual processing: *Cells and pathways within the visual system are concerned with one or another specific aspect of the scene being viewed.*

But examine *Figure 37*. More is going on with the ganglion cells than simple responses to changes in illumination. Indeed, a cell's response depends on where a spot of light falls within the cell's receptive field. Centrally the cells respond with either an ON or an OFF response, but when the light is positioned in surrounding regions, the opposite response is elicited: ON-center cells have an OFF surround, and OFF-center cells have an ON surround. Furthermore, the center and surround regions inhibit one another, so when both regions are illuminated simultaneously, the response is weak.

The antagonism between the receptive field's center and surround regions can be accounted for by a reciprocal lateral inhibition in the outer retina, similar to the reciprocal lateral inhibition between optic nerve axons in the horseshoe crab eye. In the vertebrate retina, an inhibitory neuron, called the horizontal cell, mediates this inhibition in the retina's outer region. Horizontal cells extend processes widely in the outer retina. They receive input from photoreceptors and make inhibitory synapses on nearby photoreceptors and second-order cells (bipolar cells) that carry the visual message from the photoreceptors to ganglion cells (*Figure 38*). So ganglion cells with center-surround antagonistic receptive fields tell the rest of the brain not only whether light is increasing or decreasing in intensity, but also something about the distribution of light on the retina. Furthermore, the simple lateral inhibition in the outer retina can explain Mach bands, the neurally mediated enhancement of edges and borders that plays a crucial role in form detection discussed in Chapter 4. Finally, we see the beginning of color processing in the outer region of many retinas; that is, the centers of the receptive fields of certain ganglion cells in animals with good color vision are maximally responsive to light of one color—that is, they receive their input from just one cone type—whereas the receptive field surrounds are most responsive to light of another color. Thus, a lot of processing goes on in the outer retina, and this is just the beginning!

In the inner retina, between the second- and third-order cells are neurons that detect movement (*Figure 38*). When a spot of light is shone on the retina and left there, these neurons, called *amacrine cells*, initially respond vigorously, but then their responses rapidly fade away. When the light is turned off, these cells also respond, but again the responses quickly fade away. By contrast, moving the spot around on

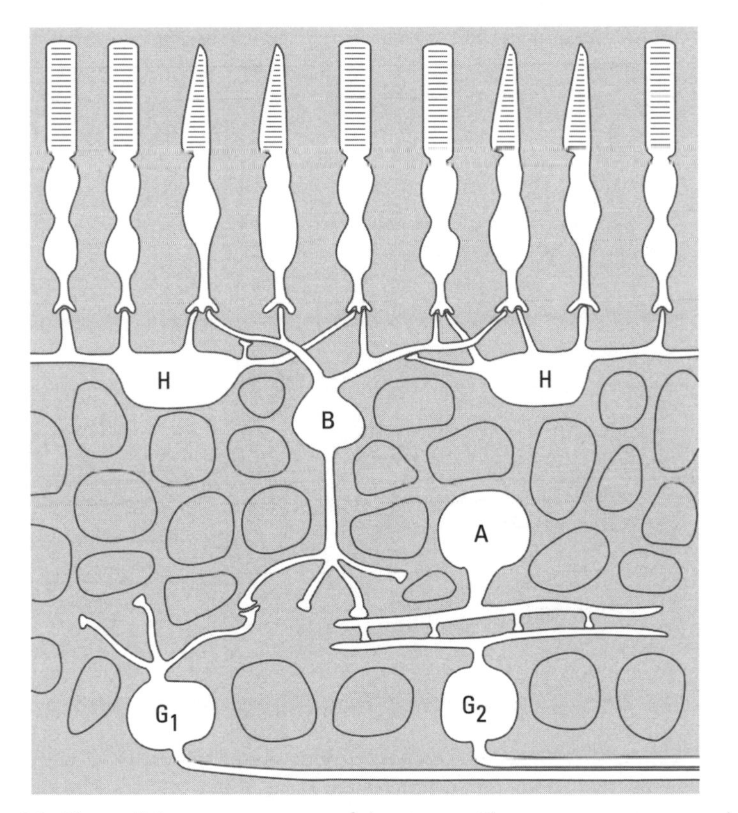

Figure 38 The cellular organization of the retina. Photoreceptors activate horizontal (H) and bipolar (B) cells in the outer part of the retina. Bipolar cells carry the visual signal from the outer to the inner retina, whereas the horizontal cells mediate lateral inhibition in the outer retina. The bipolar cells activate ganglion (G) and amacrine (A) cells in the inner retina. Some ganglion cells receive much of their input directly from bipolar cells (G_1 cell, left), whereas others receive most of their input from amacrine cells (G_2 cell, right). The ganglion cell axons run along the inner surface of the retina (bottom).

the retina elicits vigorous activity in the amacrine cells that lasts as long as the spot remains in the receptive field (*Figure 39a*). This movement sensitivity is passed on to a second group of ganglion cells, the ON-OFF ganglion cells which are highly movement sensitive and some are even direction sensitive (*Figure 39b*). A spot of light moving across the retina in one direction elicits activity, but the same light moving in the opposite direction excites nothing.

So from the eye to the rest of the brain go two messages—one

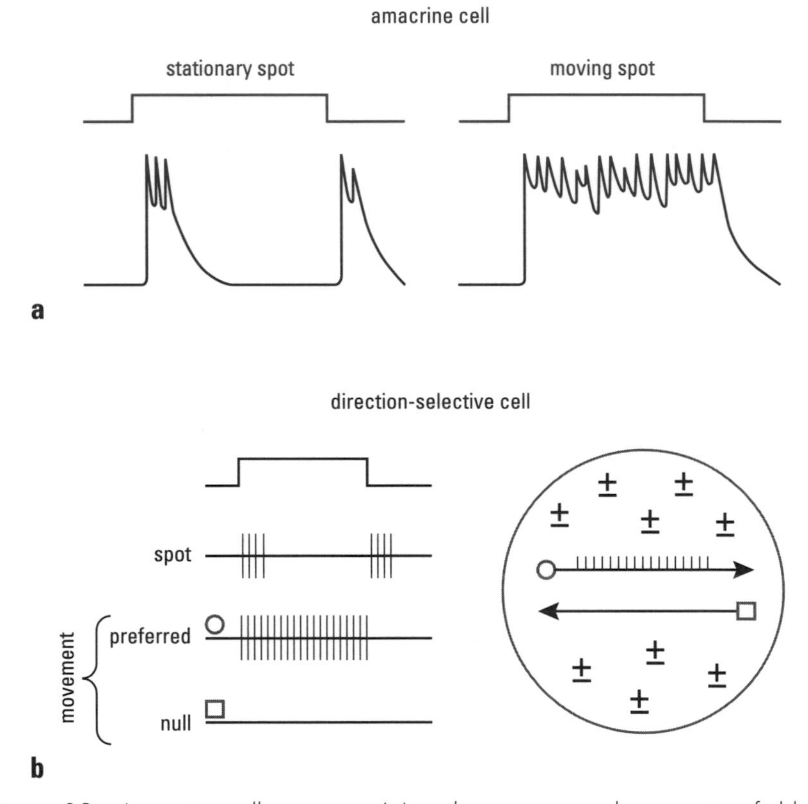

Figure 39 Amacrine cell responses (*a*) and responses and a receptive field map of a direction-selective ganglion cell (*b*). To a stationary spot of light, amacrine cells respond with a transient response at the onset and offset of the light. To a moving spot, the cell responds continuously. A directionally selective ganglion cell responds with an ON-OFF response to a stationary spot of light projected anywhere in the receptive field, indicated on the map by a ± symbol. When a spot of light is moved through the field in the preferred direction, vigorous activity is elicited in the cell. When the spot moves in the null direction, no actvity is elicited.

reflecting outer retinal processing, the other inner retinal processing—by way of two sets of cells. One set (G_1 in *Figure 38*) provides luminance, spatial, and (sometimes) color information; the other (G_2 in *Figure 38*) provides temporal information about the light impinging on the retina. The latter cells respond vigorously to movement and the visual image's dynamic properties, which leads to the second recurrent theme in visual and brain processing: *Information is processed along parallel pathways, and abundant parallel processing occurs within the brain.* So, different aspects of the visual image, such as form, color, and movement, are analyzed simultaneously along different pathways. This is a key feature of how the brain works and why our brains are far superior to any computer so far devised. We can process many things simultaneously that require neural computations: talking, walking, seeing, hearing, touching, and so forth. Computers process information serially for the most part; they do one computation at a time. Parallel-processing computors are beginning to emerge, but such computers are extremely difficult to program. Thus, even though the processing components that make up our brains, the neurons, operate slowly compared with the electronic units in computers, our brain far outstrips any computer so far devised because of our ability to parallel process in a massive way.

Later Stages of Visual Processing: The Cortex

From the retina, the visual message goes to the cerebral cortex by a way station in the thalamus, the lateral geniculate nucleus (*Figure 40*). The receptive fields of lateral geniculate nucleus neurons are very similar to those of the ganglion cells, indicating that relatively little processing of visual information occurs there. Axons from the lateral geniculate nucleus then project to the occipital lobes of the cortex, where the visual signal is first processed at the back of the occipital cortex in an area called V1 (V is for "visual").

When neurons in area V1 are recorded, and stimuli presented to the retina, a considerable elaboration of receptive field organization happens. Rather than responding well to round spots of light, either stationary or moving, presented to the retina, the cortical cells require more complex stimuli to respond maximally (*Figure 41*). In area V1, several types of neurons are recorded. *Simple cells* are found closest to the input areas in the cortex. They respond best to bars or edges of light that are oriented at a specific angle. Some cells respond best when the bar or edge is straight up or down, but others when the bar or edge is on a slant, at 45 degrees, for example. Still other cells respond best to horizontal bars or edges. Altering the optimal orientation by about 10 degrees causes the cell to fire less well.

Complex cells are found farther away from the input areas of the cortex. They also respond best to highly oriented bars of light, but the bar *must* be moving, and moving in a specific direction—at right angles to the orientation of the bar (*Figure 41*). These cells respond only weakly to a stationary bar of light, even when the orientation is correct. More specialized complex cells are also recorded and they are located even more distally from the input areas. Some of these cells have direction-selective properties. Bars moving in one direction vigorously excite the cells, but bars moving in the opposite direction inhibit the cells. Stationary bars of light projected on the retina cause little or no response, as do bars moving at right angles to the preferred direction of move-

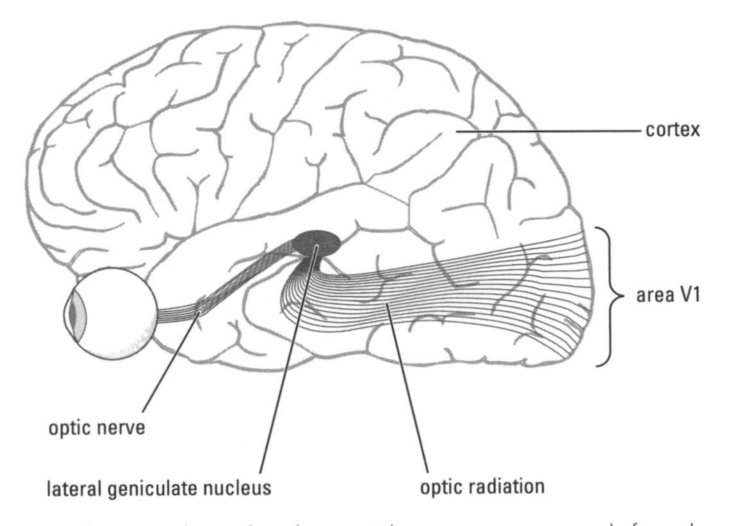

Figure 40 The central visual pathways. The optic nerve extends from the eye to the lateral geniculate nucleus in the thalamus. From the lateral geniculate nucleus, the visual message goes to the primary visual area (V1) of the cortex via the optic radiation, which consists of the axons of lateral geniculate neurons.

ment. Other specialized complex cells require moving bars that are restricted in their length. Bars that are longer than optimal begin to inhibit the cells' activity.

The main conclusion drawn from these observations is that the farther along the visual system one records a neuron, the more specific must be the stimulus presented to the retina to drive the cell maxi-

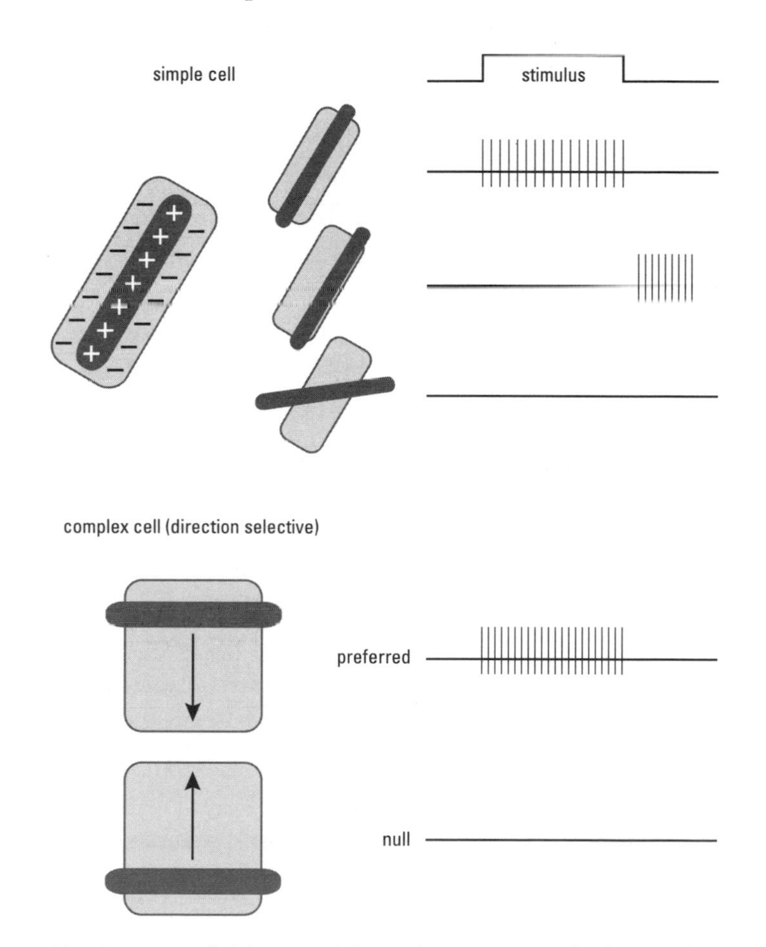

Figure 41 Receptive field maps (left) and responses (right) for simple (top) and complex (bottom) cortical neurons. The simple cell responds best to an oriented bar of light that fits its central excitatory zone (+ symbols). Moving the bar into the surround region (– symbols) elicits an OFF response from the cell. Stimulating the field with an inappropriately oriented bar of light results in little or no response.

The complex cell responds to an oriented bar of light moving at right angles to the bar's orientation. The cell illustrated is direction-sensitive. Movement in the preferred direction elicits vigorous activity; in the null direction, no activity.

mally. Whereas spots of light are sufficient to drive retinal ganglion cells vigorously, simple cortical cells require oriented bars or edges of light. Complex cells require not only oriented light bars but bars that are moving, and specialized complex cells require an oriented bar moving in one direction and/or restricted in length. You can think of this as an abstraction process; specific cells respond only when a specific set of criteria with regard to a stimulus is met. Thus, when a complicated figure is projected onto the retina, the responding cortical cells are few compared to the total number of neurons receiving input from that part of the retina—only cells respond whose receptive field requirements are matched by a part of the figure. Put in another way, components of a figure are encoded by specific neurons. A figure is thus analyzed by the visual system, with individual neurons responding to one or another part of the figure.

From cortical area V1, most visual information projects to the surrounding cortical region, area V2 (*Figure 42*). The cells in area V2 respond best to more subtle aspects of visual stimuli than do the neurons in V1, but we begin to see prominently another significant feature of visual and brain processing. That is, in area V2, there is a clear segregation of processing into separate regions. Color information is processed in a separate area from movement, which is processed separately from form. In area V2, thick and thin bands alternate, and within the individual bands color, form, and motion are separately processed. Beyond area V2, entirely separate brain areas handle aspects of the visual image. Color, for example, is processed primarily in area V4, movement in area V5, and form in area V3.

Are there consequences to this segregation? To some extent, yes. Because color is processed separately from form, it is hard to see form in an image without luminance differences; that is, when the colors are all of the same brightness in a picture, form is difficult to discern. Seeing images requires different light intensities—different colors won't do.

Another consequence of this segregation is that a stroke or other small brain injury can impair the ability to see color if the stroke or injury is limited to area V4 or another color-specific area. All other aspects of vision may be fine. The patient Jonathan I. described at the beginning of this chapter is such an example. He most likely had a permanent lesion in area V4 because of his stroke. Other patients have been described who have lost the ability to see movement, but have normal color and form vision. They may have a lesion in area V5. Loss of movement vision is severely debilitating; such patients cannot cross a street unaided because they cannot tell whether cars are moving toward them. Pouring a cup of tea from a pot is also extremely difficult. The liquid flowing from the pot looks solid to these individuals, and they cannot tell when to stop pouring because they do not see the tea rising in the cup.

Beyond areas V3, 4, and 5 are other areas, perhaps as many as twenty-five, analyzing one or another aspect of the visual image. These areas participate in two major pathways of visual information flow. One pathway progresses from the occipital lobe dorsally into the parietal lobe. Known as the *where* pathway, it analyzes such things as where an object is, or orients a subject or animal toward an object. The other pathway flows from the occipital lobe ventrally into the temporal lobe. It is known as the *what* pathway; it identifies objects. One such highly specialized area, found under the temporal lobes, performs face recognition. Patients with lesions there fail to recognize individuals visually; they may even fail to recognize a spouse by sight. Yet, when someone

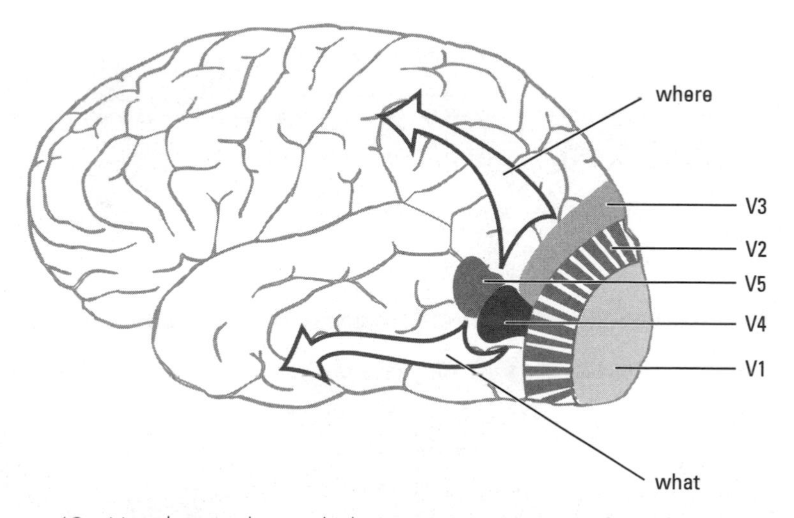

Figure 42 Visual areas beyond VI. Area V2 is segregated into bands where different aspects of the visual image are processed. Areas beyond V2 are concerned primarily with processing one of these aspects of the visual image—V3, form; V4, color; and V5, movement. Visual information thereafter flows along two cortical pathways labeled the *where* and *what* pathways.

they know well begins to speak, they instantly recognize the person; hence, theirs is clearly only a visual recognition defect.

There is something special about face recognition that we don't understand very well. For example, it is very difficult for any of us to recognize a face upside down, even of someone we know well. If recognition is made, it is usually because of a distinctive hair style or glasses rather than a recognition by features. When a face is right side up, however, recognition is easily accomplished. A dramatic illustration of the special nature of face recognition is shown in *Figure 43*. These pictures, prepared several years ago of a then well-known political figure, look quite similar when upside down. When viewed right side up, they are grossly different!

Visual Perception

What conclusions can we begin to draw about how we see and perceive objects? At least three. The first generalization has already been

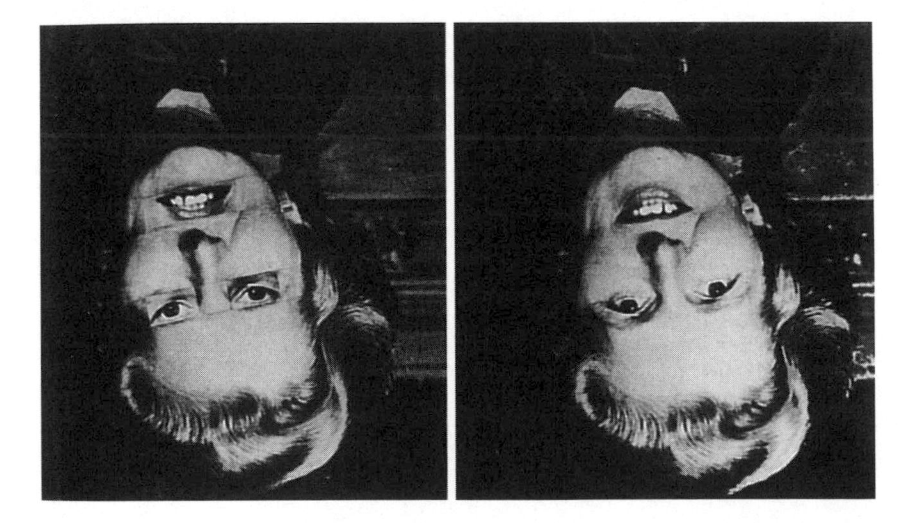

Figure 43 Upside-down pictures of Margaret Thatcher, former Prime Minister of England. The two pictures appear approximately equivalent when viewed upside down, but when turned right side up they appear very different. In one of the photographs, the eyes and mouth have been inverted.

emphasized: Individual neurons and neural pathways process one or another aspect of the visual image. Thus, massive parallel processing happens with visual information—initially in the same brain regions, eventually in distinct brain areas. A significant conclusion is that brain function has considerable localization; color is processed in a separate region from movement, which is processed separately from form.

The second generalization is that the visual system is not designed to make absolute judgments but rather to make comparisons. This is illustrated by processing in the outer retina, which results in the output (bipolar) cells having an antagonistic center-surround receptive field organization. The response of a bipolar cell depends not only on the intensity of a spot of light falling on the center of its receptive field but also on the light falling on surrounding regions. Thus, it is not the absolute intensity of light from an object that determines its lightness or darkness, but the intensity of light from the object *relative* to surrounding areas. Two practical examples illustrate this. When turned off, a television screen appears gray. When turned on, the television picture displays good blacks as well as all shades of gray as well as bright white. There is no such thing as negative light; thus the natural gray of the TV screen appears black when the set is on because of lighter adjacent areas. Reading a newspaper in very dim or very bright light is another example. In either case, the print is black and the rest of the paper white regardless of the illumination level one is in. But if you measure the light reflecting off the black print in bright light, it can exceed the intensity of light reflecting off the white areas in dim light!

Not only does perceived brightness depend on the intensity of light that surrounds an object, the object's perceived color also depends on the color of surrounding objects. No one picks out upholstery material for curtains, chair, or couch without taking a swatch home to see if it looks right in the designated space. Surround is even critical for perceived judgments of size. *Figure 44* is a picture of two women. They appear of comparable size in the picture on the left, but are vastly different in the one on the right simply because of the inappropriate position of one of the women in the right-hand picture.

This latter example brings us to the third important generalization about visual perception—it is reconstructive and creative. The image that falls on each retina is two-dimensional, yet we live in a three

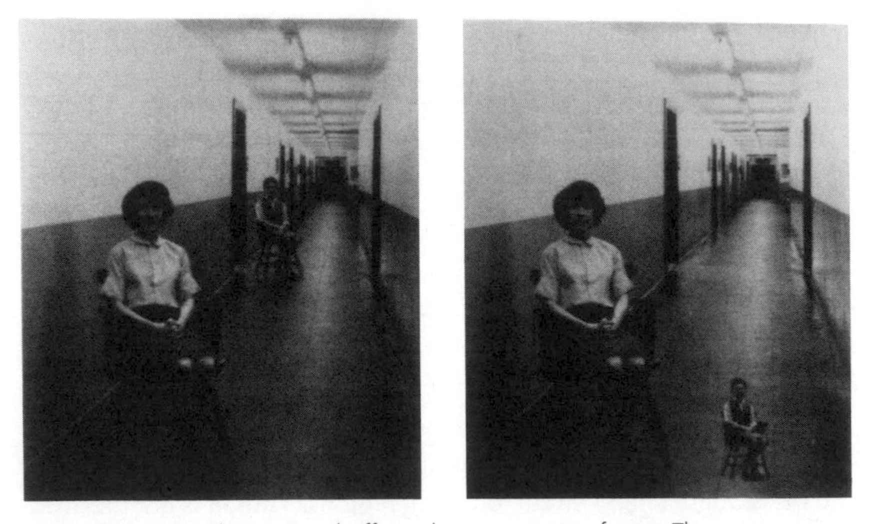

Figure 44 How the surround affects the perception of size. The scene was photographed twice—once with both women present, once with only the woman in the foreground. The image of the other woman was cut out from a print of the first picture and pasted onto the second picture. She now appears much smaller than in the first picture, but she is exactly the same size in both photographs. Measure for yourself.

dimensional world. Our visual system uses many pieces of information to reconstruct a scene. If all the pieces are not there, or are not consistent, the system tries to provide a complete and coherent picture. Not only does the visual system use the information impinging on the retinas, but it draws on visual memories and experience to construct a coherent view of the world.

The visual system can be fooled or confounded, as visual illusions show. *Figure 45a* presents classic examples of visual illusions that make it seem that lines are present when they are not, that a figure is present when it is not, and that different grades of whiteness exist when there are none. If visual information is ambiguous, we see one thing or another, but usually never both simultaneously, as shown by the familiar face-vase illusion in *Figure 45b*. The visual information coming into our eyes is invariably imperfect, yet we form in our brains a logical and complete visual world. Each of our retinas has a hole in it—where the optic nerve exits. But when we look out at the world we do not have

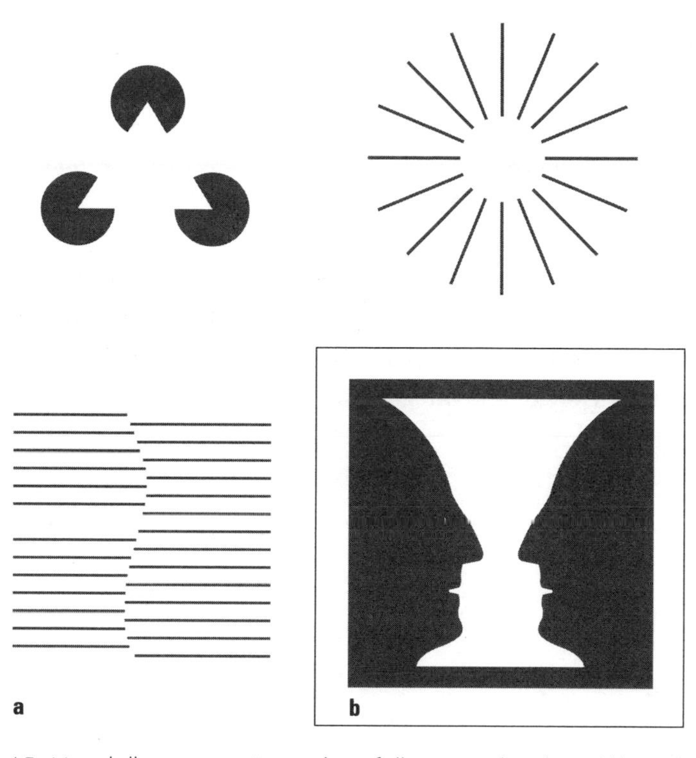

Figure 45 Visual illusions. *a*: Examples of illusionary borders. Although the shapes are not drawn as such, a well-defined triangle and circle appear in the top figures, which appear brighter than the adjacent white areas. In the lower figure, a curved line separating the two halves of the figure appears to be visible. *b*: The face-vase illusion. At any one time two black faces or a white vase are perceived. Never are the two percepts seen simultaneously.

a hole in our field of view. The brain fills in that defect so that we don't usually notice it. Yet this blind spot can be easily demonstrated, as *Figure 46* shows.

How all of what I have described above comes together to create our visual world is not yet clear. We create images in our brains by using visual information coming in, by making assumptions about shapes, colors, and movements within the scene, and by drawing on visual memories. We construct a coherent image, but not necessarily the one that is out there. Other systems in the brain operate in a similar fashion. For example, memories are often reconstructive, reflect-

Figure 46 The blind spot. Close your left eye and focus your right eye on the X. Move the page slowly toward your face and the face on the right will disappear. It reappears when the book is within a few inches of your face.

ing the confluence of a number of past experiences. Our visual system is not faultless, nor is any other part of the brain, but an important generalization is that our brains strive to give us a logical, consistent, and coherent view of the outside world.

CHAPTER 7

Development and Brain

The years 1890 and 1891 were periods of intense labour and of most gratifying rewards. Encouraged by the applause of Kölliker and convinced that I had finally found my proper path, I attacked my work with positive fury. The only explanation is that I was desirous of carrying conviction by the overwhelming volumes of my communications. During 1890 alone, I published fourteen monographs, without counting the translations. Today I am astonished by that devouring activity, which took aback even the German investigators, who are the most industrious and patient on the globe. My tasks began at nine o'clock in the morning and usually continued until about midnight. And the strangest thing is that the work gave me pleasure. It was a delicious rapture, an irresistible enchantment. . . .

In my inmost heart, I regard as the best of my work at that time the observations devoted to *neurogeny*, that is to embryonic development of the nervous system. I may be pardoned if, despite my promise of brevity, I point out a few of their antecedents.

"Since the silver chromate [method] yields more instructive and more constant pictures in embryos than in the adult, why" I asked myself, "should I not explore how the nerve cell develops its form and complexity by degrees, from its germinal phase without processes [branches], as [Wilhelm] His demonstrated, to its adult or definitive condition? In this developmental course, will there not, perhaps, be revealed something like an echo or recapitulation of the dramatic history lived through by the neuron in its millennial progress through the animal series?"

With this thought in mind, I took the work in hand, first in chick embryos and later in those of mammals. And I had the satisfaction of discovering the first changes in the neuron, from the timid efforts at the

> formation of processes [branches], frequently altered and even resorbed, up to the definitive organization of the axon and dendrites. . . .
>
> I had the good fortune to behold for the first time that fantastic ending of the growing axon. In my sections of the three-day chick embryo, this ending appeared as a concentration of protoplasm of conical form, endowed with amoeboid movements. It could be compared to a living battering-ram, soft and flexible, which advances, pushing aside mechanically the obstacles which it finds in its way, until it reaches the area of its peripheral distribution. This curious terminal club, I christened the *growth cone*.
>
> —Excerpted from Santiago Ramón y Cajal,
> *Recollections of My Life* (Cambridge, Mass.: MIT Press, 1989)

One of biology's greatest challenges is to understand how the brain forms. Indeed, how the brain and brain cells develop has long fascinated those studying the brain, as the above excerpts from Santiago Ramón y Cajal's autobiography make evident. From a few undifferentiated cells in the young embryo, all of the neurons and glial cells that make up the brain arise. The brain consists of hundreds of areas, each carrying out a specific task. Many areas possess neurons unique to those specific parts of the brain. And within each area, the neurons connect with one another, and some project to other areas often considerable distances away. How does all of this happen? What tells an undifferentiated cell to become one kind of a neuron or another? How do axons find their way to distant targets? How do neurons know which cells to synapse on, and whether to form excitatory, inhibitory, or modulatory junctions?

Of the estimated 100,000 genes present in the human genome, the brain probably utilizes 50,000 of them; and of these, 30,000 may be unique to the brain. The brain is clearly our most complicated organ, and so far we are ignorant about how its development is orchestrated. In humans, virtually all of our neurons form before birth. If we assume that the brain contains 100 billion cells (a conservative estimate), then at least 250,000 neurons are generated per minute *in the nine-month gestation period. Even this astonishing figure is an understatement, because cell generation is not constant over the entire gestation period. Indeed, many more than 250,000 neurons per minute are often being formed.*

A long debate has festered over how much environmental factors affect brain development and maturation. How much can brain development be affected by early experience and training? Although virtually all of our neurons are present by birth, and probably no significant amount of neuronal cell division takes place in the brain after about six months of age, it takes many years for the brain to become fully mature. How hard-wired is the brain at birth? How malleable is it thereafter? How much of the final product is due to nature, how much to nurture? This chapter examines these vital questions.

The brain begins to take shape in humans about three weeks after conception. A group of cells, about 125,000 in number, form a flat sheet along the dorsal (back) side of the embryo. All of the neurons and glial cells of the nervous system derive from these cells, known as the *neural plate.*

Between the third and fourth week of development, the neural plate folds inward and creates a groove that eventually closes into a long tube, the *neural tube.* All of the central nervous system derives from the neural tube; the anterior part becomes the brain proper and the posterior part becomes the spinal cord. By about forty days of development three swellings along the anterior part of the neural tube become apparent; these eventually form the forebrain, midbrain, and hindbrain.

As the neural tube forms, some cells on either side are left behind. These cells, known as *neural crest* cells, come to lie on either side of the neural tube. Much of the peripheral nervous system derives from the neural crest cells, including sensory neurons, Schwann cells, cells of the autonomic nervous system, certain cranial nerves, and so on. *Figure 47* presents a schematic view of how the neural tube and neural crests form, and *Figure 48* depicts the development of the human brain from thirty days after conception, or from shortly after the neural tube has formed.

What causes the formation of the neural plate? Covering the early embryo are *ectodermal* cells that eventually create skin. Lining the inside of the embryo are *endodermal* cells that will form the stomach, intestine, and other internal organs. At about two and a half weeks of development, a third, intermediate layer of cells forms, the *mesoderm.* The mesoderm differentiates into many tissues, including muscles, skeleton, and cardiovascular system; it turns out that the mesoderm is responsible for generating the neural plate cells. As the mesoderm forms, it slides between the ectoderm and endoderm. As it slides under ectodermal cells on the dorsal surface of the embryo, it induces these cells to change their fate—to become neural plate cells.

Experiments with salamanders demonstrated that neural plate cells are induced from ectodermal cells by underlying mesodermal cells. In the 1920s the German biologist Hans Spemann and his student Hilde Mangold showed that if clumps of mesodermal cells are transplanted from one part of the embryo to another, the transplanted cells will induce overlying ectodermal cells to become neural tube cells regardless of where on the embryo surface those cells reside. So, for example, a second neural plate can be induced to form on the ventral side of the embryo if the transplant is placed there. These seminal experiments evoked much interest and prompted more research. Scientists showed that if mesodermal cells are prevented from migrating under the dorsal ectodermal, no neural plate develops and the embryo fails

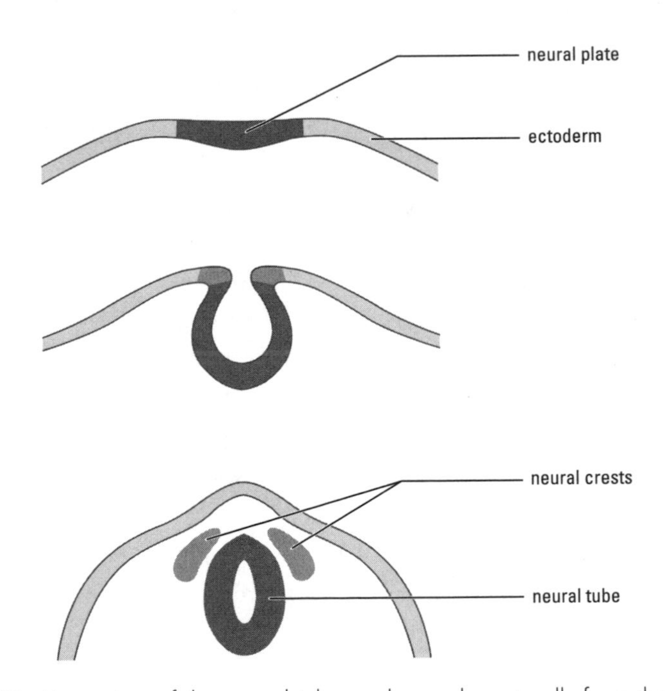

Figure 47 Formation of the neural tube and neural crest cells from the neural plate. During the third and fourth weeks of development of the human embryo, the neural plate invaginates, forming the neural tube. The neural tube differentiates into the central nervous system—the brain and spinal cord. The neural crest cells are derived from neural plate cells positioned laterally along the plate and left behind during the formation of the neural tube. Neural crest cells differentiate into the peripheral nervous system.

to form a nervous system. This finding confirmed the notion put forward by Spemann and Mangold that the neural plate cells are induced by mesoderm. Hilda Mangold was tragically killed in a kitchen explosion at the time the first experiments were published. Spemann, whose brilliant career in developmental biology extended from the turn of the century to the end of the 1930s, was awarded a Nobel Prize in 1935, primarily for his work on induction.

How can mesodermal cells induce ectodermal cells to become neural plate cells? An early and obvious suggestion was that mesodermal

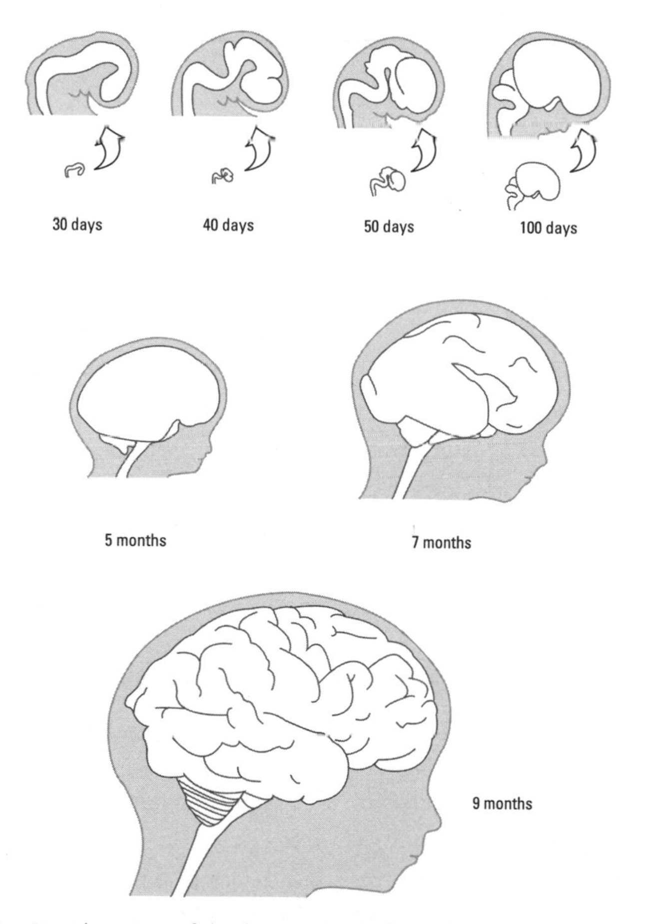

Figure 48 Development of the human brain from the neural tube. The detailed drawings at the top are enlarged relative to the middle and bottom drawings. The smaller drawings at the top indicate the actual sizes of the developing brain at early stages relative to those at latter stages. Note that the infolding of the brain's surface occurs rather late in development.

cells release chemical substances that induce ectodermal cells to change their fate. Several experiments support this idea. For example, when pieces of ectoderm are cultured in the presence of mesoderm, they become neural plate cells—but not when mesodermal cells are absent. Furthermore, by placing porous filters between mesoderm and ectoderm, one can define the sizes of the agents that mediate the induction. These experiments suggest that the inducing molecules are small proteins. Yet identifying the precise molecules has been thorny.

Becoming a Neuron

Two theories attempt to explain neuronal differentiation, and they reverberate through much of the thinking about brain development and maturation. The first, the *cell lineage theory*, maintains that cells inherit developmental characteristics and are not strongly influenced by their environment during development. A cell's lineage determines its fate. The second theory, called the *induction theory*, contends that a cell in the nervous system differentiates according to its position in the embryo during development and to clues from nearby cells or the environment. The lineage theory holds that precursor cells are committed to specific fates dictated by inherited directives; the induction theory proposes that precursor cells can become one of several cell types depending on the position-dependent signals they encounter during development. One developmental biologist describes the cell lineage theory as the European plan (what you become depends on your ancestors) and the induction theory as the American plan (what you become depends on your neighbors). Although in simpler invertebrates lineage theory accounts for much of neural development, in more complex animals, including all vertebrates, development appears to be indeterminate. What a precursor cell becomes depends mainly on its location in the embryo during development.

How might environment and cellular interactions govern cell fate in the developing nervous system? The developing eye of the fruit fly has provided some of the most compelling insights into the mechanisms. The fruit fly's eye is like that of the horseshoe crab—a mass of hundreds of photoreceptor units or ommatidia (*see Figure 21*). In the fruit fly each ommatidium contains eight photoreceptor cells that are precisely arranged in the structure. Each of the photoreceptor cells can be identified, and they are labeled R1 to R8 (R represents "retinular," the technical name for these photoreceptor cells).

During development of the ommatidium, the R8 cell differentiates first, followed by the R2 and R6 cells, which differentiate simultaneously. Then the R3 and R4 cells differentiate, followed by the R1 and R5 cells. The R7 cell, which contains a visual pigment that absorbs in the ultraviolet region of the spectrum, is the last to form.

This strict sequence of development suggests that the earlier cells are responsible for the differentiation of the later cells; indeed, if the developmental sequence is disturbed, the ommatidium does not form properly. A mutation in fruit flies in which the R7 photoreceptor does not form at all—discovered because the flies are insensitive to ultraviolet light—elucidates the nature of the signaling mechanisms and intracellular pathways. The mutant is called *sevenless*. The mutated gene in normal flies codes for a sizable protein that extends across the cell membrane. On the outside of the cell, the protein is receptor-like; on the cell's inside, it has a kinase-like structure. (Recall that kinases are enzymes that add phosphate groups to proteins, and thus activate or inactivate them.) Presumably an extracellular signal binds to the receptor and activates the intracellular kinase. By phosphorylating intracellular proteins, the kinase initiates a biochemical cascade within the cell leading to differentiation. If the cascade is not initiated, the cell does not differentiate into a photoreceptor cell. Indeed, what we observe in sevenless flies is that the precursor cell destined to become the R7 photoreceptor in a normal eye becomes a nonneural cell in the mutant eye.

What can be said about the signal that interacts with the sevenless gene protein and the downstream intracellular proteins phosphorylated by it? Progress on both fronts has been made, although the story is not yet complete. A second fruit fly mutation, which also hinders a developing ommatidium to form an R7 cell, has provided clues to the intercellular signal and its origins. This mutant, called *bride of sevenless* or *boss*, affects the R8 cell. The defective gene in boss mutants codes for a membrane protein in normal flies. The guess is that the portion of this protein on the outside of the R8 cell activates the sevenless receptor protein. Thus, the R7 photoreceptor forms when the R8 photoreceptor cell provides a signal to a precursor cell. Since the R8 signal is a membrane protein, this implies that in ommatidial development direct cell-cell contacts are required to induce differentiation of R7. *Figure 49* depicts schematically the interaction between R7 and R8.

What about downstream pathways after the receptor has been activated? Several downstream pathway proteins have been identified, and many are kinases themselves. Thus, after the R7 receptor-kinase protein is activated, other kinases are too. Targets of at least one of these kinases are proteins that serve as *transcription factors*. Such factors alter the expression of genes within cells by binding directly to regions of DNA in the nucleus that regulate the turning on or the turning off of

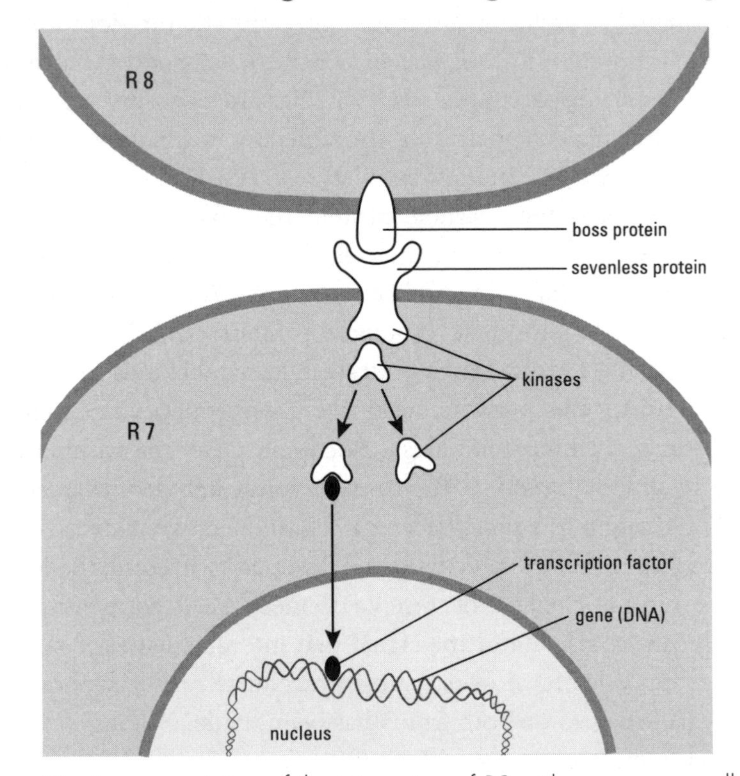

Figure 49 A representation of the interaction of R8 with a precursor cell that leads to the formation of the R7 photoreceptor. A protein (boss) on the R8 cell binds to the sevenless protein on the precursor cell, resulting in the activation of the kinase associated with the sevenless protein. Activation of the sevenless kinase leads to the activation of other kinases and ultimately to the activation of transcription factors that regulate gene expression in the nucleus.

genes. The idea is this: when the sevenless receptor-kinase protein is activated, appropriate genes are turned on in the precursor cell, which cause it to differentiate into the R7 photoreceptor cell.

In other systems it is likely that diffusible substances control cell differentiation, but the same principles as described above apply. Sometimes the signaling molecules are small proteins called *growth factors*. These proteins activate membrane receptors linked to a cascade of intracellular kinases, ultimately turning on or off specific genes. Thus, the general sequence of events in the developing fruit fly eye is probably true for the differentiation of neurons and glial cells throughout much of the brain.

How Do Axons Find Their Way?

Once neurons begin to differentiate, they extend out branches, both dendrites and axons, as noted in the Cajal excerpts at the chapter beginning. This leads eventually to the formation of synapses between neurons and, ultimately, to the wiring of the brain. How do neurons know which cells to synapse upon, and how do the axons of neurons find their way? Sometimes axons must travel substantial distances to their targets.

Again chemical signaling has long been implicated as playing a critical role in cell-cell recognition. The proposal is that as neurons differentiate they become chemically specified; they make specific proteins on their surfaces that enable other neurons to recognize them.

Early experiments that supported this chemoaffinity hypothesis were performed in the early 1940s by Roger Sperry at the University of Chicago. Sperry studied the projection of retinal ganglion cell axons to the tectum in cold-blooded vertebrates such as fish and frogs. Such projections are quite orderly; ganglion cell axons from one part of the retina project to a particular region of the tectum. Such projections are called *topographic*—they are accurate, consistent, and invariant from one animal to another—hence a retinal map is impressed on the tectum. The right retina projects to the left tectum, and vice versa; and the tectal map is inverted relative to the retinal map. Thus, when a specific region of the retina is activated, a certain region of the tectum responds. *Figure 50* shows schematically the retinal-tectal projections in goldfish and how the retina maps onto the tectum.

As pointed out in Chapter 5, central nervous system axons regenerate in cold-blooded vertebrates after they have been severed. So, for example, if the optic nerve is cut in a fish or frog, the optic nerve will regenerate and reform synaptic connections in the tectum. With this reinnervation, vision is restored to the animal. What Sperry did was to cut the optic nerve in a newt (a frog-like animal), rotate the eye 180 degrees in the socket, and then reattach the eye. He observed that after the optic nerves had regenerated, the animal could see but its visual world was upside down and inverted from right to left! So when animals so altered were feeding, they consistently misdirected their

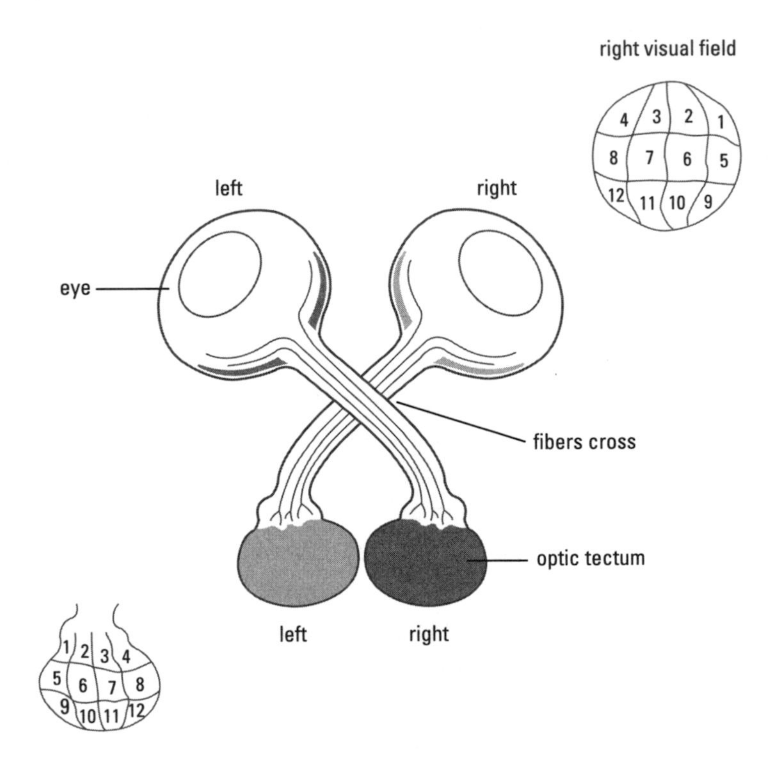

Figure 50 Retinal-tectal projections in the goldfish. The ganglion cells from the left retina project to the right tectum and vice versa (left). The projection is orderly and consistent. Ganglion cell axons from one retinal region end up in a specific part of the tectum, as shown in the maps. Corresponding areas between the retina and tectum are indicated by the numbers.

approaches to food by 180 degrees. If a fly they wished to capture was up and to the right, they moved down and to the left.

These experiments indicated that the severed optic nerve axons had grown back to their original location and vision had been restored. But since the animals' eyes were inverted, they saw an inverted world and responded in this way. Over time, there was no recovery; the animals were permanently altered. The conclusion drawn was that optic nerve axons can recognize the cells they are intended to synapse upon; the cells have complementary markers that allow for a mutual recognition. Many experiments have since been carried out supporting this general idea. However, the bulk of the experiments do not support the notion that a specific retinal axon is wired to a specific tectal cell. Rather, the view today is that axons know in a general sense where they are to go—to which area of the brain and where in that area they should make synapses. But there is some flexibility during both development and regeneration; initial synapses may be broken and new ones formed during development, maturation, and normal functioning of the brain.

Axons grow via specialized structures, called *growth cones*, that are flattened expansions of the tips of growing axons from which extend numerous fine branches (*Figure 51*). Growth cones were first observed by Ramón y Cajal, and a wonderful and graphic description of these structures is included in the excerpts from his autobiography at the beginning of the chapter. While axons grow, the growth cone is in constant motion, extending and retracting its fine processes and exploring the surrounding area. As the growth cone moves along, it adds new membrane to the axon, and hence it lengthens.

The rate and direction of growth cone movement depend on several factors: the substrate on which the growth cone is moving, the presence of chemicals in the environment, and the electrical fields in the axon's vicinity. Substrate texture and adhesiveness can affect growth cone movement as well as recognition molecules present in the substrate. Diffusible molecules in the environment can also affect growth cone movement—both positively and negatively. Thus, some substances encourage growth cone movement while others inhibit it. Axons growing long distances can also be guided by specialized cells called *guidepost neurons* found at intermediate distances along the way.

The guess is that the guidepost neurons secrete an attractant chemical sensed by the growth cone. Axons grow toward the guidepost neuron but do not stop when they reach the cell. Rather, after making contact with a guidepost neuron, they move on to the next guidepost cell. If guidepost cells are damaged, axonal pathfinding may be disrupted.

Initially axon pathfinding in the brain occurs early in development when distances between structures are much shorter than they are in the adult. So another explanation of how axons can find their way over long distances is that the early axons serve as *pioneers*. Early in development a few axons from any one region find their way to their targets, and, as the brain grows, the axons simply stretch. Axons from neurons that differentiate later find their way by growing along the pioneer axons.

When axons reach their targets, they then make synaptic connections with appropriate cells—the ones they recognize. Synapse formation often requires reciprocal interaction. Substances released from the growth cone, including neurotransmitters, neuromodulators, and

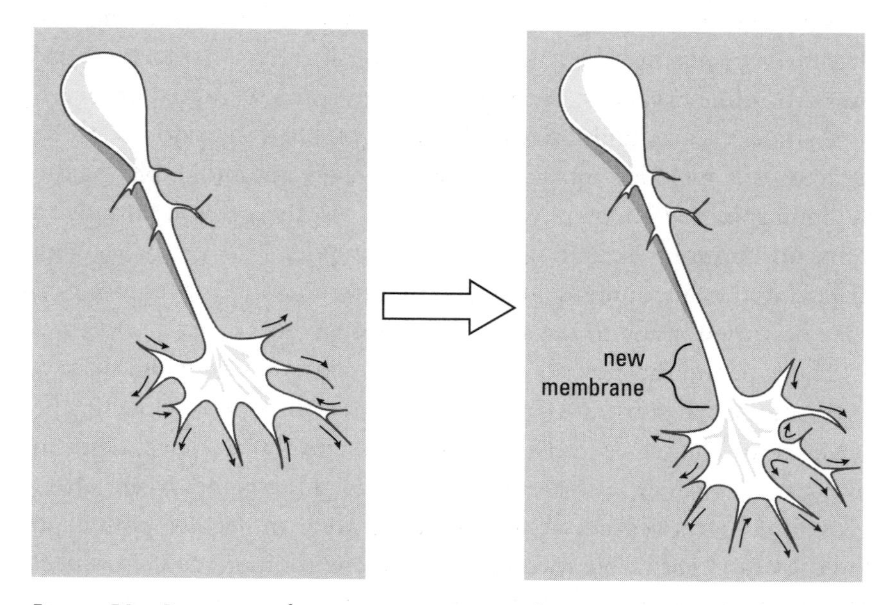

Figure 51 Drawings of an axon growing out from a neuron and its growth cone. The growth cone is continually extending out fine branches that explore the surrounding area. The axon grows by adding new membranes adjacent to the growth cone.

other molecules, initiate postsynaptic structures to form. And the postsynaptic element also releases substances that induce the growth cone to develop into a mature presynaptic terminal.

The Maturing Brain

A most surprising feature of brain development is that many neurons die during the maturation process. In many brain regions, 30 to 75 percent of the neurons die during development. Why? Much of the cell death appears to relate to a competition for target cells and synaptic sites. Neurons that successfully form synapses with target cells survive, while neurons that lose in the competition for a target cell die. If target cells are eliminated—by removing part of the target, for example—neuronal cell death increases.

Survival of presynaptic neurons depends, therefore, on signals from target cells. Again, these are believed to be chemical signals. One such signal is a protein termed *nerve growth factor* (NGF), which has been extensively studied by Rita Levi-Montalchini, an Italian scientist working at Washington University in St. Louis and in Rome, who discovered the substance. She found that if excess NGF is made available during development, presynaptic cell death diminishes, and if an antibody that inactivates NGF is given to an animal during brain development, neuronal cell death is excessive. She and others have shown that NGF does more than promote survival of neurons. When given to young animals, it can enhance the number and extent of dendrites and multiply the synapses made by those neurons. Moreover, NGF can promote axonal growth. For the discovery of NGF and its significance, Levi-Montalchini was awarded the Nobel Prize in 1986.

Another key feature of the brain maturational process is that once a synapse forms, it is not necessarily permanent. Indeed, during brain development, axons typically have numerous terminals, and they make more synapses than they do in the fully mature brain. During brain maturation, therefore, axon terminal fields and synaptic contacts often are rearranged to become more restricted. This means that neurons establish appropriate connections when they develop, but during maturation the connections are rearranged to provide more precise wiring found in the adult.

The big questions are how long the plasticity of neuronal structure and connectivity lasts, and how it is influenced by experience and environment. For certain regions of the brain, especially ones concerned with higher neural processing, substantial neuronal modification and synaptic plasticity can continue throughout life. We learn and remember new things throughout life, and the mechanisms for memory and learning involve alterations in neuronal and synaptic processes (see Chapter 8).

The extent of neural plasticity is under intensive investigation. What we know unequivocally is that synaptic wiring of the primary visual cortex can change drastically when visual input is altered. Yet these alterations can be induced only in young animals or humans. Such experiments on animals, carried out primarily by Torsten Wiesel and David Hubel at the Harvard Medical School, have had a profound impact on our understanding and thinking about brain maturation. Their experiments provided some of the first evidence that substantial remodeling of brain circuitry can occur. Furthermore, their experiments have relevance with regard to phenomena such as language acquisition; the same principles appear to apply. For their groundbreaking work on the primary visual cortex, Wiesel and Hubel were awarded the Nobel Prize in 1981.

Visual System Development and Deprivation

How mature is the visual system at birth in a visually inexperienced animal? Are the cells wired up correctly at birth? The answer is simple. Recordings from newborn monkeys and cats indicate that the responses of neurons in the visual cortex are surprisingly adult-like in their behavior. Many cells are less vigorous in responding to visual stimuli than are neurons in older animals, and a few neurons fail to respond at all to visual stimuli. But neurons with good orientation selectivity are evident, and they may have simple, complex, or specialized complex receptive field properties quite like those seen in the adult cat or monkey. The conclusion is that at birth, the visual cortex in a cat or monkey is wired up correctly and the neurons respond like adult neurons. Thus, visual experience is not necessary for establishing the complex wiring of the cortical neurons, and by inference, also the wiring of retinal and lateral geniculate nucleus neurons. The system is ready to go at birth.

If, though, a young monkey or cat is deprived of form vision in one or both eyes, severe alterations in vision can develop. Such a situation can happen in young humans if, for example, the lens of the eye is clouded—has a cataract. In monkeys or cats, a similar situation can be induced by sewing an eyelid shut, or by applying a light diffuser to the eye. In all of these situations, light can reach the photoreceptors, so the deficit is not the result of light deprivation. Rather, it arises because sharp images cannot form on the photoreceptor array; only very fuzzy images reach the retina.

In humans and animals, visual acuity is severely reduced in such conditions. If the deprivation is monocular, visual acuity measured in the deprived eye is sharply depressed whereas visual acuity in the other eye is fine. If both eyes are deprived, visual acuity measured in either eye is decreased. This loss of visual acuity is termed *amblyopia*.

To understand the changes in the visual system after deprivation, Wiesel and Hubel recorded from neurons in the primary visual cortex of cats and monkeys that had one eyelid closed at birth or shortly thereafter. The recordings were of animals four months old or older. Before the experiments were undertaken, the closed eye was opened so the two eyes could be stimulated equivalently. However, virtually all the cells recorded in the cortex received input only from the eye that had been open from birth. The few cells driven by stimulation of the deprived eye usually responded abnormally.

When cells in the retina or lateral geniculate nucleus (LGN) were recorded, the responses were quite normal. This means that the bulk of the changes were in the cortex, which underwent substantial modification, but the structures providing input to the cortex were relatively unaffected by the deprivation.

What is going on in the cortex? The physiological studies described above suggest that input from the open eye occupied a disproportionate amount of the cortex compared to the closed eye. In normal animals, the two eyes have equal representation in the cortex; when recording from a population of cells in the primary visual cortex, most neurons are preferentially driven by one eye or the other, but both eyes have equal cortical representation. The cells that receive the bulk of their input from one eye are clustered in columns or stripes, about 0.5 mm in thickness, that run across the cortex (*see Figure 52a*), but the same amount of cortex is devoted to input from each eye. The stripes alternate so one stripe has cells that are primarily driven by the right eye, the next stripe by cells from the left eye, and so forth. In form-deprived animals, this balance is dramatically altered; the open eye takes over territory belonging to the closed eye.

This striking fact can be illustrated anatomically by injecting a radioactive amino acid into one eye and looking at the pattern of radioactivity in the cortex. How this works is as follows. The radioactive amino acid is taken up by ganglion cells in the injected eye. The amino acid is transported along the ganglion cell axons to the LGN. Some of the labeled amino acid is released from the ganglion cell axon terminals, taken up by the LGN neurons, and transported to the cortex by way of their axons. This takes about a week, at which time the

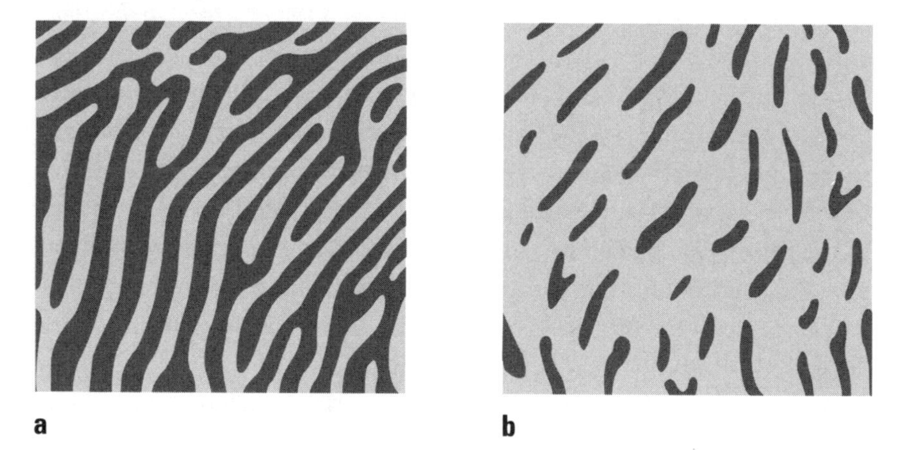

Figure 52 *a*: A representation of the ocular dominance columns or stripes that extend across the primary visual cortex. The cells in one stripe receive their input preferentially from one eye and the stripes alternate. Each eye has equal representation in the cortex in the normal animal. *b*: In an animal in which form vision has been deprived in one eye, the amount of cortex receiving input from that eye is much reduced. The ocular dominance columns or stripes are thin and discontinuous.

terminals of LGN axons that received input from the injected eye are radioactive. A flat section through the cortex, placed on a piece of photographic film, demonstrates the distribution of the labeled terminals, because radioactivity, like light, exposes silver grains in film.

As noted above, input from the two eyes in a normal cortex is organized in columns or stripes that extend across the cortex, and the thickness of the stripes is about equal for the two eyes. In a monocularly deprived animal, however, the stripes representing input from the closed eye are much smaller and even discontinuous, as shown in *Figure 52b*. How does this happen? One idea is that lateral geniculate axons compete for cortical space and synaptic connections in the young animal. So long as each eye provides equivalent input to the cortex, both eyes have equal cortical representation. If one eye has less input or a defective input, the other eye dominates the competition and ends up with more cortical area and can make more synapses.

Can one induce such changes to occur in the visual cortex throughout life? The answer is unambiguously no. In adult animals (cats and monkeys) and in humans, form deprivation does not have dramatic effects on either visual acuity or the responses of cortical neurons. Lid closure or the presence of a cataract for months to years does not cause amblyopia in an adult cat, monkey, or human. Such changes require that the deprivation take place soon after birth. The period after birth when such changes can be induced is called the *critical period*. In monkeys, the critical period is between birth and one year of age, with deprivation for the first six weeks causing more severe change than deprivation later. In humans, the critical period does not begin until about six months of age but extends to five or six years. Deprivation need not be long during the early part of the critical period for severe changes to happen. A few days of deprivation in the first two weeks of a monkey's life can make changes as severe as the ones in *Figure 52b*.

How easily can the changes be reversed? Surprisingly, they are not readily reversed, even in young animals. If in a young animal a closed eyelid is opened after a short period of deprivation, little recovery is observed after months to years of the eye remaining open. A trick learned long ago by ophthalmologists in treating children who are amblyopic because of crossed eyes can lead to some recovery. A patch covering the good eye for a part of every day forces the child to use only the amblyopic eye. Gradual recovery of acuity occurs in the bad eye of the child, as with animals rendered amblyopic by monocular form deprivation.

The amblyopia from crossed eyes in children is similar to the amblyopia induced by monocular form deprivation. In these children, one eye, usually the straighter eye, gradually becomes dominant, and visual input from the other eye is ignored. The ignored eye becomes highly amblyopic. Such an amblyopia can be induced in animals by altering the eye muscles so their eyes are crossed. Cortical changes in these animals closely resemble the changes after monocular form deprivation. Again, the critical period when changes can be induced matches the form-deprivation critical period.

Fundamental alterations in the brain can thus be induced by altering sensory experience or input. The young brain is much more susceptible to these environmental effects than is the more mature brain. Furthermore, some areas of the brain may show striking changes (i.e., visual cortex) whereas other regions (retina and lateral geniculate nucleus) are minimally changed by alterations in sensory input.

Learning Languages

Is there evidence for similar phenomena in other parts of the brain and for other systems? The answer is unequivocally yes, and language acquisition is one good example. Everyone knows it is much more difficult to learn a new language as an adult than as a child (or, for that matter, to learn to ride a bicycle or play the piano). Linguists talk of a critical period for language acquisition that extends from about the age of two up to puberty. After puberty the ability to learn a new language is dramatically reduced and is not particularly age-dependent. Learning a new language at forty is little different from learning one at twenty. Some individuals have an easier time learning a new language as an adult than do others, and some individuals become quite proficient in a language learned as an adult, yet most linguists agree that the accent for a language learned as an adult is never perfect and that late learners can easily be detected by a language expert.

Why do humans lose the ability to learn a new language perfectly as they mature? Young children are sensitive to a broad range of sounds but lose the ability to distinguish certain ones unless they continue to hear them. Most Japanese people, for example, cannot distinguish an *r* from an *l* sound, although Japanese infants can readily discriminate these sounds. Conversely, a young child can easily imitate almost any sound an adult makes, but as the youngster grows up, this ability is compromised if he or she does not continue to hear the sounds. We brought our daughter to Japan at the age of five for a seven-month period and in just four months she became fluent in Japanese. Her pronunciation of Japanese words was generally superior to that of my wife, who had been studying the language intensively for the previous three years! After we left Japan, my daughter stopped speaking Japanese and declined to speak it with her mother despite our urging her to do so. When she does return again to that country or attempts to relearn the language, I wonder how that short but early experience will affect her ability to understand and speak it.

One idea about what goes on in the brain's language areas during the critical period is that synapses rearrange and gain or lose territory depending on language experience, much as happens in the visual cortex after monocular form deprivation. At birth or shortly thereafter, it is likely that the circuits have formed to make and discriminate all possible language sounds, but if the circuits are not exercised they are rearranged or even lost.

Experiments on animals' visual systems may again enlighten us. Virtually all neurons in the visual cortex have a high degree of orientation selectivity; any neuron responds best to a bar or edge that has a precise orientation. All orientations are represented in about equal numbers when a group of neurons are recorded randomly. If, though, the visual stimuli provided an animal during the critical period are biased so only vertically oriented lines, edges, and objects are present, subsequent recordings from groups of neurons indicate that most of the neurons respond only to vertically oriented stimuli and few, if any, to horizontally oriented ones. This also happens with horizontal stimuli; the neurons respond preferentially to horizontally oriented bars or edges.

The conclusion is that the circuitry necessary to carry out complex neural tasks is formed during the brain's development. But at least some of this circuitry is labile, which means that it has to be used if it is going to be retained. The old adage "Use it or lose it" fits perfectly here. Synapses may be lost, and brain territory can be taken over by nearby neurons if synapses are not active and the circuitry in a piece of brain tissue not used. These alterations happen most readily in the young animal or human, but even adults have neural plasticity. How much is not known for certain. What is unequivocal is that the brain can be shaped considerably by early sensory experience. Knowing this, we should take pains to enrich youngsters' environments. A simple example is to provide recordings that have the sounds of all languages. These might help them later in life if they attempt to master a second language.

CHAPTER 8

Language, Memory, and the Human Brain

By the 1920s it was thought that no corner of the earth fit for human habitation had remained unexplored. New Guinea, the world's second largest island, was no exception. The European missionaries, planters, and administrators clung to its coastal lowlands, convinced that no one could live in the treacherous mountain range that ran in a solid line down the middle of the island. But the mountains visible from each coast in fact belonged to two ranges, not one, and between them was a temperate plateau crossed by many fertile valleys. A million Stone Age people lived in those highlands, isolated from the rest of the world for forty thousand years. The veil would not be lifted until gold was discovered in a tributary of one of the main rivers. The ensuing gold rush attracted Michael Leahy, a footloose Australian prospector, who on May 26, 1930, set out to explore the mountains with a fellow prospector and a group of indigenous lowland people hired as carriers. After scaling the heights, Leahy was amazed to see grassy open country on the other side. By nightfall his amazement turned to alarm, because there were points of light in the distance, obvious signs that the valley was populated. After a sleepless night in which Leahy and his party loaded their weapons and assembled a crude bomb, they made their first contact with the highlanders. The astonishment was mutual. Leahy wrote in his diary:

> It was a relief when the [natives] came in sight, the men . . . in front, armed with bows and arrows, the women behind bringing stalks of sugarcane. When he saw the women, Ewunga told me at once that there would be no fight. We waved to them to come on, which they did cautiously, stopping every few yards to look us

> over. When a few of them finally got up courage to approach, we could see that they were utterly thunderstruck by our appearance. When I took off my hat, those nearest to me backed away in terror. One old chap came forward gingerly with open mouth, and touched me to see if I was real. Then he knelt down, and rubbed his hands over my bare legs, possibly to find if they were painted, and grabbed me around the knees and hugged them, rubbing his bushy head against me. . . . The women and children gradually got courage to approach also, and presently the camp was swarming with the lot of them, all running about and jabbering at once, pointing to . . . everything that was new to them.
>
> The "jabbering" was language—an unfamiliar language, one of eight hundred different ones that would be discovered among the isolated highlanders right up through the 1960s. Leahy's first contact repeated a scene that must have taken place hundreds of times in human history, whenever one people first encountered another. All of them, as far as we know, already had language. Every Hottentot, every Eskimo, every Yanomamo. No mute tribe has ever been discovered, and there is no record that a region has served as a "cradle" of language from which it spread to previously languageless groups.
>
> As in every other case, the language spoken by Leahy's hosts turned out to be no mere jabber but a medium that could express abstract concepts, invisible entities, and complex trains of reasoning.
>
> —Excerpted from Steven Pinker, *The Language Instinct*

Paleontologists tell us that modern humans appeared about 200,000 years ago. The skulls of early humans were much like ours, which suggests that their brains were similar to ours. So it is likely that they possessed mental faculties not very different from our own and, in particular, that they used language. The origins of language almost certainly go back even further. Our human ancestors of two and a half to two million years ago made and used crude stone tools, and by one million years ago, they were crafting hand axes and using fire. It is almost certain that these ancient people used some language communication.

It is generally agreed that what most distinguishes humans from animals is language. And, as the excerpt above points out, the universality of language among peoples is striking. All humans possess it, and furthermore, as Steven

Pinker puts it in The Language Instinct, *"There are Stone Age societies but there is no such thing as a Stone Age Language." All spoken languages are amazingly sophisticated and complex. Although considerable efforts have been made to teach language, both oral and sign, to animals, especially chimpanzees, virtually everyone agrees that no animal has ever acquired anything comparable to human language. And it is not simply the ability to speak that makes the difference. Sign languages developed by many deaf communities are as complex and sophisticated as spoken languages. The ancestors of humans and chimpanzees diverged about five to seven million years ago. Thus language had at least this much time to evolve. What evolutionary pressures led to language development in our ancestors but not in the ancestors of other modern primates is a matter of speculation.*

Language

Speech, reading, and writing are controlled within the cortex, and two cortical areas are crucial for speaking and for understanding speech or writing. One area is located in the frontal lobe, close to the region of the primary motor area involved in initiating face, tongue, and jaw movements. This area, discovered by Pierre Paul Broca, a nineteenth-century French neurologist-anthropologist, is known as *Broca's area.* Broca studied the brains of patients who had lost the ability to speak—they were aphasic—and found in every case a specific area in the frontal cortex that was damaged (*Figure 53*).

We now know that Broca's area is concerned mainly with the production of speech. Patients with lesions in Broca's area know what they want to say, but their ability to articulate what they wish to say is impaired. They cannot form the sounds of words properly. With severe lesions in Broca's area, no speech sounds or only rudimentary ones may be made. The first patient Broca studied was called Tan, because this was the only speech sound he could make—"tan-tan-tan" was all he uttered. Lesions in Broca's area also lead to deficits in writing or in sign language; hence Broca's area is involved in more aspects of language than just word articulation. Linguists agree that it is also important for grammatical processing in general.

The other language area was found in the temporal lobe by Carl Wernicke, a German psychiatrist. Wernicke's area sits between the primary auditory and visual areas and is concerned primarily with the comprehension of speech and with reading and writing (*Figure 53*). Patients with lesions in Wernicke's area can articulate words normally, but their word choice is often inappropriate. The words they say are understandable but when strung together are often unintelligible. Wernicke's area is also concerned with reading and writing. Lesions that disrupt visual input to Wernicke's area result in patients that have difficulty reading; patients with lesions in Wernicke's area closer to the primary auditory cortex often have difficulty in understanding speech.

A curious feature of Broca's and Wernicke's areas is that they are most often located only on one side of the brain, typically the left side. In about 95 percent of the population these areas are located in the left hemisphere. In just 5 percent, they are located in the right hemisphere. Somewhat more left-handed individuals (about 20 percent) have the language areas on the right side, but even in left-handers, the majority (about 70 percent) have their language areas in the left hemisphere.

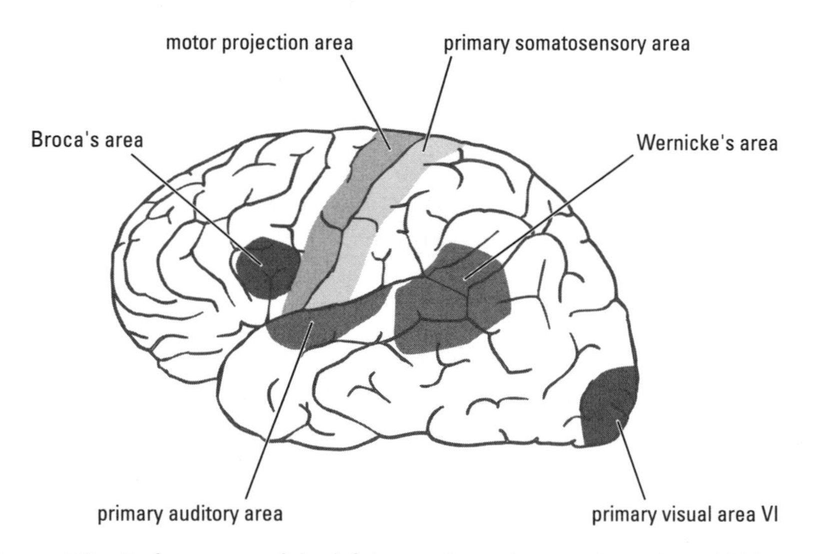

Figure 53 Surface view of the left hemisphere showing Broca's and Wernicke's areas. Broca's area is close to the region of the primary motor area concerned with face, tongue, and jaw movements. Wernicke's area is between the primary auditory and visual areas.

(About 10 percent of the left-handed population have language areas in both hemispheres.)

In right-handed individuals, then, about 97 percent of the population have language areas located on the left side. What goes on in comparable areas on the brain's right side? Again, clues from the clinic provide insights. Individuals who have lesions in Broca's area and who are highly aphasic often sing very well, which implies that the ability to make music is localized in the right hemisphere, in an area corresponding to Broca's area. Individuals with lesions in the right hemisphere, in the region corresponding to Broca's area, may lose the ability to sing or to play a musical instrument, but they speak perfectly normally.

Why are speech areas confined to just one hemisphere? No one knows. Are there consequences of having the speech areas mainly in the left hemisphere? For normal individuals, the consequences are minimal. Psychologists have shown that subjects will recognize words a bit more accurately and quickly when they are flashed into their right visual fields as compared to the left. (Remember that sensory information on the right side of the body is processed by the left side of the brain.) So something is lost when information is transmitted between the two hemispheres. But the general consensus is that information exchange between the hemispheres, which is mediated by a massive band of axons connecting the two (the *corpus callosum*, depicted in *Figure 54*), is very efficient.

In certain individuals, the corpus callosum is cut surgically because of a tumor in the area or incurable epilepsy. In these individuals, clear deficits can be demonstrated. While these individuals can readily name objects placed in the right hand or describe events seen in the right visual field, they cannot verbalize what they are touching or seeing when the object or events are on the left side or in the left visual field. They are aware of the objects or events but cannot describe them. Their perception of objects and events has been disconnected from their language centers.

Are other faculties localized to just one hemisphere? None that we know of; representation in both hemispheres is the general rule. Yet certain functions seem better done by one hemisphere or the other. So, for example, in addition to speech, reading, and writing, arithmetic calculations and complex voluntary movements may be mediated pri-

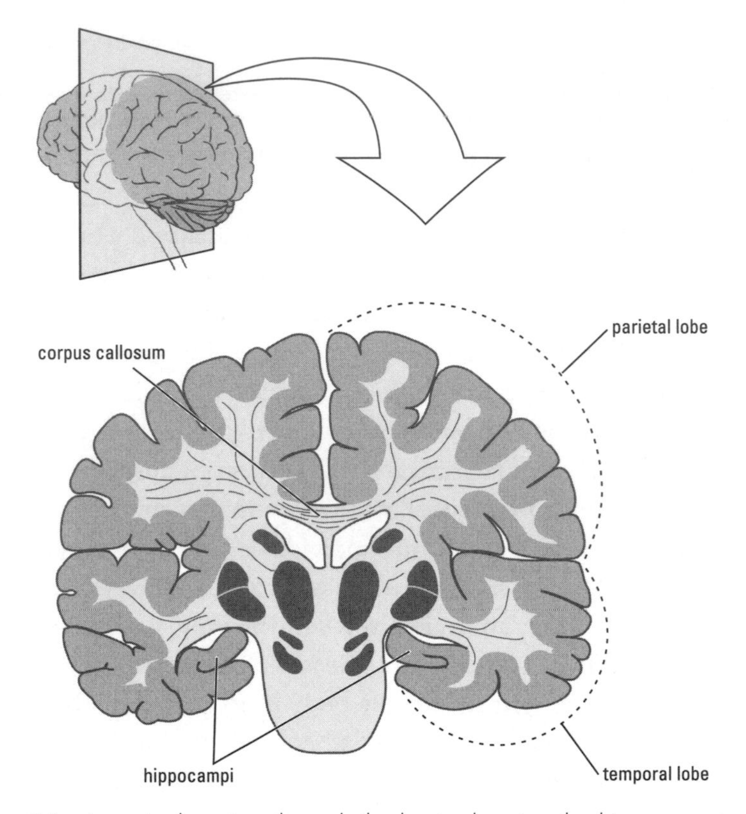

Figure 54 A vertical section through the brain showing the hippocampi tucked medially behind the temporal lobes. This section is identical in position to the one shown in Figure 32.

marily in the left hemisphere. For right-handed individuals, skilled movements of the right hand are superior to those of the left hand, and undoubtedly this reflects left cortical dominance. Conversely, in left-handed individuals the right cortex would have dominance.

The right hemisphere in most individuals seems concerned more with complex visual, auditory, and tactile pattern recognition. Spatial sense, intuition, and singing and music making are primarily right-hemisphere functions in most individuals. Are individuals better in one or another of these attributes because one hemisphere is more developed or dominant? Although this is an intriguing hypothesis, and undoubtedly has some truth to it, concrete evidence in support of the hypothesis is lacking.

Exploring the Human Brain

The human brain is unique in its language ability. It undoubtedly has other unique abilities, not as well defined or understood, that underlie our intelligence and rich mental lives. How can we understand the human brain better? We cannot often record from neurons in human brain tissue or carry out experiments on humans as we do on animals. But the clinic provides a wealth of information, and in this book I have given several examples of patients with localized cortical or other brain lesions who have exhibited neurological deficits. Broca's area was discovered in this way, as were several other cortical areas.

Are there other ways we can gain information about the human brain? One method that has proved useful stems from a surgical procedure to treat *epilepsy*, developed in the late 1930s by Wilder Penfield, a Canadian neurosurgeon. Epileptic seizures occur when a population of neurons in the brain becomes spontaneously active and stimulates abnormal neural activity to sweep across the cortex. This causes an involuntary and frequently dramatic change in behavior, often accompanied by loss of consciousness. Epilepsy is usually caused by a small group of diseased or damaged cells that initiate the sweep of synchronous brain activity. The purpose of the surgery is to remove the abnormal cells.

The surgical procedure attempts to remove as little brain tissue as possible. To accomplish this, Penfield took advantage of the fact that no pain fibers are in the brain itself; the brain can be touched or even cut and a patient feels nothing, yet neurons can be stimulated electrically from the brain surface. So with a small electrical probe Penfield stimulated discrete areas of the cortex in search of the abnormal cells that lead to a seizure. The brain is exposed by removing the skull overlying the region suspected of harboring the abnormal cells. This part of the procedure is done while the patient's head and skull are anesthetized locally and the patient is asleep. Once the cortex is exposed, the patient is kept fully awake. During the main part of the operation, the surgeon converses with the patient and records sensations, movements, feelings, and even emotions. The surgeon can stimulate the normal cortex in search of abnormal cells, and thus has an opportunity to see what effects the stimulation of the cortex can have on a patient. Some of Penfield's results were very startling.

Stimulating primary sensory areas evoked sensations appropriate to the area being stimulated. For example, electrical stimulation of the primary visual area in the occipital lobe at the back of the brain usually induced a visual response—a flash or streak of light was reported by the patient. Stimulating the auditory areas on the sides of the brain (temporal lobes) might cause the patient to hear a sound, while stimulation of the primary somatosensory areas in the parietal lobe usually induces the sensation of being touched.

Stimulation of the primary motor cortex (where fine movements are initiated) resulted usually in a small movement of the body—bending a finger or moving part of the face. These effects, which were highly reproducible, prompted Penfield to return to an area previously stimulated and evoke the same response. In doing so, he could map in detail the brain's sensory areas and primary motor area (*see Figure 34*, which is derived from Penfield's maps). A map of the visual field, for example, is on the primary visual area, whereas a representation of the body surface is along the primary somatosensory cortex. These maps are consistent and coherent, but distorted. This is clearest in maps of the primary somatosensory and motor areas, where a representation of the body is impressed on the cortical surface. But body areas that have greater touch sensitivity or finer movements—fingers, hand, face, and lips—occupy a disproportionate amount of cortex. More cortex is devoted to their control or sensation than to areas of the body that we control less well or have less sensation from, such as our feet and toes.

Perhaps the most surprising results obtained by Penfield occurred when he stimulated the more ventral regions of the temporal lobes (*Figure 54*). On several occasions, such stimuli evoked such vivid memories in patients that they believed they were reliving an experience. Not only would they remember an event in exquisite detail, but they felt emotions related to the event. The events were not necessarily major ones in the lives of the individuals. For example, a mother recalled being in her kitchen listening to her small son playing outside. She was aware of neighborhood noises, including the sounds of passing cars.

One of the best-documented examples of temporal lobe stimulation evoking a memory concerned a young woman who heard a song being played by instruments when a specific region of her cortex was stimulated. Whenever that area was stimulated she heard the same song, and the elicited music always started in the same place. The music was so clear to her that she believed it was being played by a phonograph in the operating theater. Later she wrote to Penfield describing what she heard: "There were instruments . . . as though being played by an orchestra. I actually heard it." She added, "I could remember much more of it in the operating room after hearing it than I could three or four days later. The song is not as real to me now as it was in the operating room."

How are such observations interpreted? Because these "experiential" responses were rarely elicited—only about 10 percent of temporal lobe stimulations evoked such responses—firm conclusions have been difficult to make, though it's clear that memories can be evoked by stimulating parts of the temporal cortex. Furthermore, the richness of the evoked memories suggests that memories can be stored in much more detail than we realize or can ordinarily recall. Finally, that patients felt emotions during the evoked memories means that stimulation of the cortex can bring back to consciousness an experience that happened long ago. The fact that such memories are evoked only when regions of the temporal lobe are stimulated suggests that the temporal lobes play a central role in memory storage. Indeed, we now have abundant evidence that a region of the brain tucked behind the temporal lobes, the *hippocampus* (*Figure 54*), is critical for long-term memory storage in humans and other mammals.

Retaining Memories

The crucial role of the hippocampus in the formation of long-term memories was demonstrated dramatically and tragically in 1953 by a young patient who was severely epileptic. This patient (HM) was twenty-seven years old at the time and had frequent and debilitating seizures, so he was unable to work or lead a normal life. It was believed that the hippocampi on both sides of his brain were diseased, and so both were removed surgically. (Previously, the hippocampus on one side of the brain had been removed in patients without significant effects). After removing both hippocampi, doctors discovered, much to their dismay, that HM, although cured of his epilepsy, no longer could remember events or facts for more than a few minutes. His short-term memory appeared to be unimpaired, but long-term memory mechanisms had been permanently disrupted. Memories of events prior to the operation were retained, but new experiences or facts were quickly forgotten.

HM has been studied extensively by psychologists, especially Brenda Milner, a Canadian psychologist, for over forty years, and virtually no changes have occurred in the patient's ability to remember facts or events. He retains such memories for only a few minutes. If he keeps thinking about a fact or event, he can continue to recall it for some time; once he is distracted, though, the event or fact is quickly forgotten. Even after forty years, he does not know Dr. Milner and she must reintroduce herself to him whenever they meet. But HM can learn new motor skills or routines and retain them for a long time. (Riding a bicycle is an example of a complicated learned motor skill.) Memories, then, are often classified by psychologists into two types, *declarative* and *procedural*. The former are memories of facts or events, whereas the latter are retained motor skills or routines. Declarative memories cannot be permanently stored in patients with hippocampal lesions. Procedural memories, by contrast, can be retained in such patients, and there is evidence that the cerebellum participates in the learning and retention of procedural motor skills. When patient HM was asked to perform a motor task to test if he had retained a particular skill, he would deny that he had ever done that task before. His subsequent performance showed clearly that he had retained the skill learned earlier. But even normal individuals are often unaware of the details of a learned skill. For example, when bicyclists are asked what they do when their bicycle begins to lean to the right, most say they would lean to the left. But leaning to the left would increase the right tilt. What is done is to turn the handle bars to compensate for the tilt.

Although bilateral surgical excision of the hippocampi is no longer contemplated, there are many cases of patients with degenerative diseases involving both hippocampi. These patients are typically unable to form long-term memories, confirming the role of this brain structure in long-term memory mechanisms.

As might be expected, there is enormous interest in the hippocampus among neuroscientists. What is special about this region of the brain? Does it provide clues to how the brain remembers things? The structure of the hippocampus is distinctive from that of other brain regions, but its structure does not provide any special insights into this issue. Yet a striking physiological observation made in 1973 by two scientists, Timothy Bliss and Terje Lomo, who were then working in England on the hippocampus, led to an explosion of research continuing up to the present. Bliss and Lomo discovered that following strong stimulation of nerve pathways leading into the hippocampus, postsynaptic responses of neurons in the hippocampus were potentiated. That is, the responses of hippocampal neurons to weak stimuli were increased significantly after a strong potentiating stimulus (*Figure 55*). If the strong input stimulus was repeated several times, the potentiation of the responses could be induced to last for days or even weeks. The phenomenon, known as *long-term potentiation* or LTP, indicates that long-term changes in synaptic efficacy can be induced in single hippocampal neurons by priming stimuli. LTP thus appeared to be a model for how long-term changes can be induced in the brain during memory formation.

Much has been discovered about the mechanisms underlying LTP, but all the details are still unknown. Many of the same mechanisms described in Chapter 4 that occur in the *Aplysia* nervous system during habitutation and sensitization of the gill-withdrawal reflex appear to be involved. Ca^{2+} and cyclic AMP have been implicated, as has protein phosphorylation. There is some evidence for altered release of transmitter from presynaptic terminals, similar to that in *Aplysia*, but also alterations are induced in the postsynaptic neurons. Neuromodulatory mechanisms, including the activation of second-messenger cascades, are implicated, which results in many biochemical changes in the hippocampal neurons during LTP.

In addition to long-term potentiation, the hippocampal neurons can generate *long-term depression* (LTD) in which, after priming stimuli are given, the postsynaptic responses of hippocampal neurons are depressed for days to weeks. Mechanisms like the ones involved in

LTP might account for LTD, and these results indicate that neuronal activity can be depressed as well as potentiated on a long-term basis. Some have proposed that LTD is more closely related to memory formation than is LTP, but firm evidence is lacking. There is as yet no unequivocal link between either LTP or LTD and memory formation, although by knocking out (by genetic means) one type of kinase activated by Ca^{2+}, not only is LTP blocked, but the ability of a mouse to navigate a maze is impaired. These mice exhibit a variety of deficits

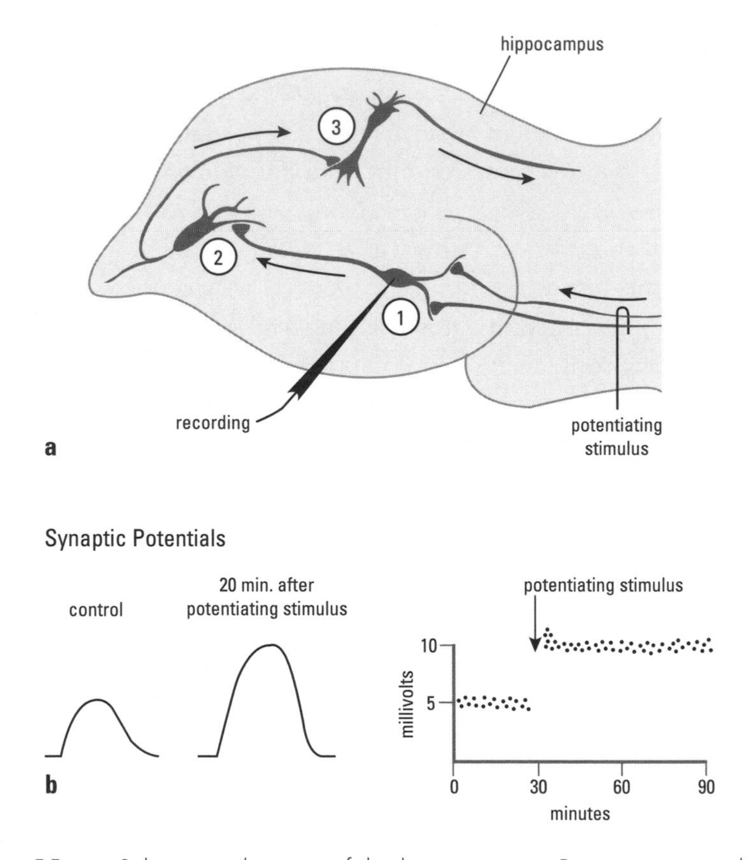

Figure 55 *a*: Schematic diagram of the hippocampus. Potentiating stimuli presented to the axons innervating the dentate neurons (1) result in long-term potentiation in these neurons. Long-term potentiation can also be induced in the CA3 (2) and CA1 (3) neurons by providing potentiating stimuli to the axons innervating these neurons. *b*: Long-term potentiation. On the left are excitatory synaptic potentials measured before and twenty minutes after a potentiating stimulus. On the right is a plot of synaptic potentials measured before and after a potentiating stimulus.

and behavioral changes, so the conclusion that this kinase is critical for LTP and memory storage is quite tentative.

The evidence that the hippocampus is essential for long-term memory is unequivocal, but how the hippocampus accomplishes this task is not yet clear. It would appear that long-term memories are not permanently stored in the hippocampus but transferred elsewhere, probably to various regions of the cortex. This idea is prompted mainly by observing patients like HM, who despite having bilateral hippocampal lesions can continue to recall events early in their lives, but not events that happened after their lesions occurred. Exactly how memories are stored in neurons or in neuronal circuits remains a mystery. The best model remains that of long-term habituation and sensitization of the gill-withdrawal reflex in *Aplysia*, in which changes in synaptic number and structure have been observed, suggesting that biochemical and gene expression changes occur in specific neurons during the formation of long-term memories.

A well-established but curious observation is that older long-term memories are more persistent in many forms of brain disease than are more recent memories. With progression of a brain disease, it is not uncommon that a patient's time reference regresses. Such regression happened with my grandfather, who in his seventies and eighties suffered from chronic diabetes. He came to live in the city where I was growing up and initially knew where he was and could correctly identify everyone in our family. As he aged, however, he moved back in time, believing my mother was his sister and I was one of his sons. Furthermore, he thought he was in the Midwest, where he grew up, and not on the East Coast with us. Eventually he identified my mother as his mother and me as a brother. Why older memories are often more resistant to brain disease than are newer memories is not known; when we discover how and where long-term memories are stored in our brains, an answer to this mystery may be forthcoming.

Brain Imaging

Today, epilepsy can be effectively controlled by drugs, and so surgical treatment of epilepsy is rarely performed. Exploring the cortex of a patient undergoing neurosurgery is also discouraged today, which limits opportunities to extend Penfield's observations. But over the past two decades, powerful noninvasive imaging techniques have studied the brain in awake—even behaving—human subjects. The techniques provide a wealth of new information on human brain structure; in particular they have shed light on the functional role of many parts of the cortex, especially the association areas.

Some techniques enable investigators to observe which parts of the brain are active when a subject is performing a neuronal task. These methods are based on the fact that when a brain region is active, blood flow to the region increases. Brain tissue requires oxygen to keep functioning; greater blood flow allows for local increases in oxygen and glucose levels necessary for maintaining the neurons. It is these increases in oxygen or glucose that are most often measured. This is accomplished in PET (positron emission tomography) scanning by giving subjects radioactively labeled oxygen or glucose, presenting a stimulus or asking the subject to perform a task, and then determining where hikes in radioactivity are happening. The radioactive isotopes are short-lived (and thus safe); the breakdown products of the radioactive atoms are detected and localized by a scanning device surrounding the head. Functional magnetic resonance imaging (fMRI), a more recently developed technique, measures the increase in blood flow directly—it does not require the injection of any substances into the subjects.

The results elegantly extend earlier findings from brain lesions studies or cortical stimulation experiments. *Figure 56* schematically depicts brain scans of a subject hearing, seeing, speaking, and generating words. The stimuli or tasks activate distinct areas of the cortex. Activity generated in an area promotes more blood flow, as reflected in the drawing by the dark shading. When the subject hears words, activity is greatest in the primary auditory cortex and also in Wernicke's area. When seeing words, activity heightens in the primary visual area in the occipital lobes and in secondary or association visual areas along the *what* pathway that extends ventrally in the cortex from the occipital to the temporal lobes. When a subject is asked to speak words that had been seen or heard, activity increases in the appropriate region of the primary motor cortex and in premotor regions important in planning and programming motor movements. A surprise was that Broca's area did not increase in activity when the subject was asked to perform this task. But when the subject was asked to generate new words based on the words he heard or saw, there was prominent activity in Broca's area (along with activity in the primary and premotor motor cortices, not indicated in *Figure 56*).

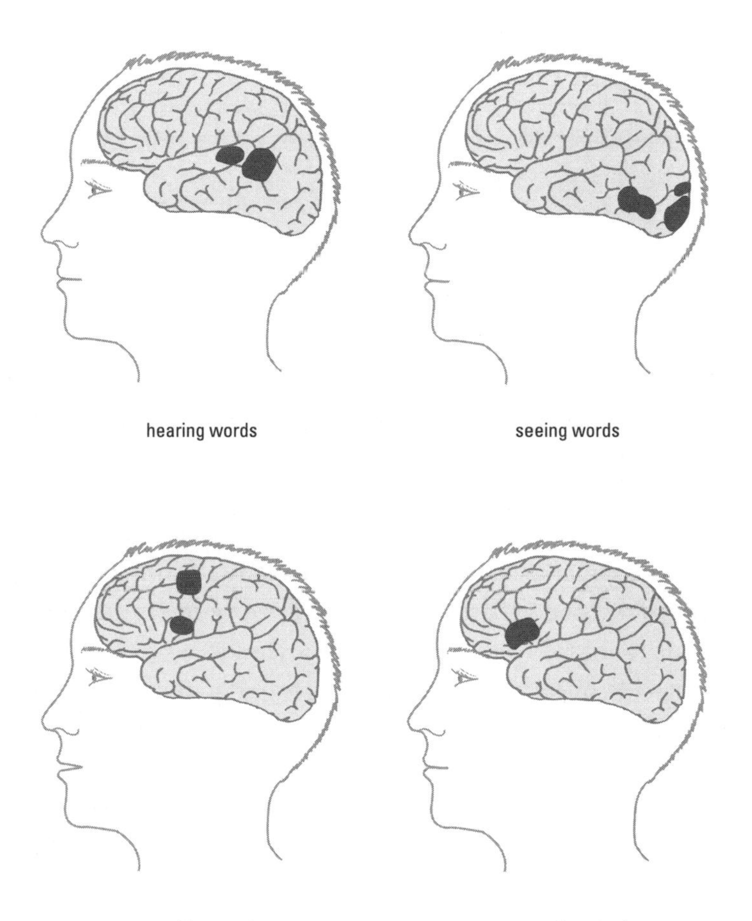

Figure 56 Schematic representations of PET scans taken when a subject is hearing, seeing, speaking, and generating words. When the subject is hearing words, activity is increased in the primary auditory and Wernicke's areas. When the subject is seeing words, activity increases in the primary visual area and in areas along the *what* pathway. Speaking words is associated with increased activity in the primary motor cortex and in certain premotor areas. Generating words results in increased activity in Broca's area.

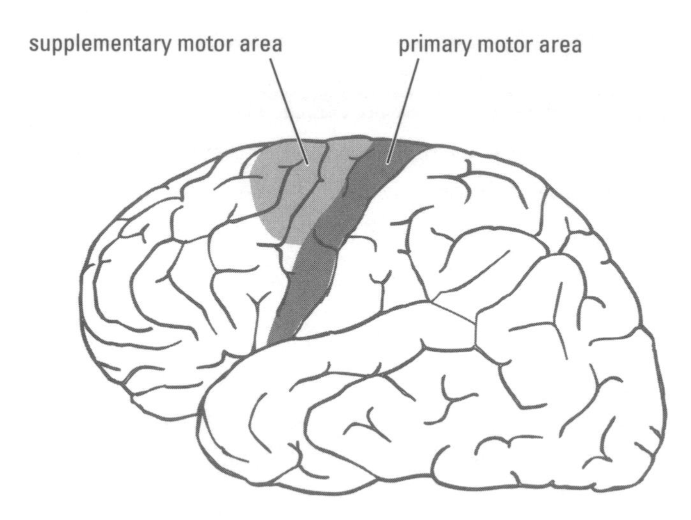

Figure 57 A surface view of the cortex showing the supplementary motor cortex, a premotor area, adjacent to the primary motor cortex.

These data, then, confirm earlier findings on the role of these brain areas, but they also provide new insights into how the brain processes information or accomplishes a motor task.

The power of these imaging techniques is yet to be realized, and their resolution is still crude. Yet important new findings are already appearing. A dramatic example was provided by a Danish scientist, Per Roland, who wanted to know how motor movements are controlled by the brain. He asked subjects to make a sequence of complex finger movements. He observed, as expected, greater activity in the primary motor cortex in the region where finger movement is initiated. But he also observed activity in another area in the frontal lobes, in one of the premotor areas called the supplementary motor area (*Figure 57*). It was known that the premotor areas are related to activity in the primary motor area, but their exact role was unclear. Roland next asked his subjects to mentally rehearse the complex finger movements but not move the fingers. He observed enhanced activity only in the premotor area, not in the primary motor cortex! These data suggest that the supplementary motor area helps in planning and programming motor activity, whereas the primary motor cortex is the locus for initiating skilled motor activity.

At present, many researchers are exploiting these new techniques and improving the methodology. Although it is still early to draw many firm conclusions, some fascinating and provocative observations have been reported. For instance, Stephen Kosslyn at Harvard has shown that when we form a mental image of something (think of Snoopy, for example, with his long droopy ears), the visual areas of the occipital cortex are more active, as though we are actually seeing a picture of Snoopy!

The challenge is to provide a stimulus to a subject or to require a task specific enough so one or a limited number of brain regions show increased activity. It is not easy to present a stimulus or task that allows a crisp interpretation of the result. Although it is simple to activate massive regions of the brain, it is much harder to limit activity to specific brain regions. Another problem is that when subjects are experienced with a task, brain activity patterns can change. Thus, selecting a task that remains novel to the subject upon repetition is essential. These technical issues undoubtedly will be overcome with further experimentation. As resolution improves so smaller and smaller brain regions can be visualized, our understanding of the role of brain regions and how they interact while the brain is performing tasks will increase dramatically.

CHAPTER 9

The Emotional Brain

Elliot was a good husband and father, had a good job and enviable personal, professional and social status. Regrettably, his life was soon to fall apart. He first developed severe headaches and found it difficult to concentrate. Subsequently, he appeared to lose his sense of responsibility, and his work often had to be completed or corrected by others. His physician suspected a brain tumor and the suspicion was correct.

The tumor was located in the lower part of the frontal lobes, approximately in the midline, so that both frontal lobes were being compressed. By the time the tumor was removed it was the size of a small orange. The tumor was removed, along with surrounding regions of the frontal lobes. The surgery was successful, but following the surgery, Elliot was a dramatically different person. He seemed not to have lost any intelligence, but rather his ability to get things done was severely impaired. He was unable to manage his time appropriately and used poor judgment when trying to accomplish even simple tasks. After repeated incidents and an apparent refusal to take advice and to do things correctly, he lost his job. He tried other jobs, but failed at them as well. He squandered his savings on inappropriate ventures, and soon his marriage collapsed. A second marriage also failed. Elliot appeared perfectly healthy and intellectually competent, but he was no longer an effective human being.

Elliot was aware that something was wrong, but when he talked about his life, he did so dispassionately—like an uninvolved spectator. Never did he express sadness, impatience or even frustration. Seldom did he show any anger—he was constantly calm, relaxed and detached.

> In short, he had lost his feelings—things that once evoked strong emotions no longer did so. He showed very little reaction—either positive or negative—regardless of the situation.
>
> —Adapted from Antonio R. Damasio,
> *Descartes' Error: Emotion, Reason, and the Human Brain*

I began this book by suggesting that feelings, emotions, consciousness, understanding, and creativity are all aspects of mind. Feelings and emotions—fear, sadness, anger, anxiety, pleasure, hostility, and calmness—localize to certain brain regions. Lesions in these areas can lead to profound changes in a person's emotional behavior and personality, as well as in the ability to manage one's life, as the case of Elliot described above illustrates. As for consciousness, understanding, and creativity, we cannot identify these attributes of mind with a specific brain region. In most textbooks of neuroscience or physiological psychology, these aspects of mind are not even brought up, reflecting the fact that little of real substance is known about which brain mechanisms generate them. Yet our fascination with these matters builds as we learn more about the brain.

The aspects of mind that can be explained best are the emotions and emotional behavior, and the consequences of disturbances in emotional behavior. First, as we all know, emotions invariably evoke significant bodily reactions. A frightening experience immediately causes pounding of the heart, rapid breathing, dryness of the mouth, and often sweating. Emotional experiences activate the autonomic nervous system that mediates these bodily effects. It is often difficult to disentangle the emotional reaction to a situation from the bodily effects. Indeed, William James, the well-known Harvard philosopher-psychologist, suggested at the beginning of this century that the bodily changes evoke the emotional reactions. When we encounter a frightening situation, the body's rapid response—pounding of the heart, sweating, dry mouth—leads to the emotional reaction, according to James's theory. People who receive less sensory input from the body—patients having a damaged spinal cord, for example—show more limited emotional responses. These observations appear to support James's theory. More current belief, however, is that emotions result from the combination and interaction of body and brain responses to an event.

Conscious emotions are constructed in the cortex from signals the brain receives from both the sense organs and *the internal organs. Much as the visual system reconstructs and creates an image from sensory signals (sometimes ambiguous) coming into the visual system, so do parts of the cortex create an emotional reaction from sensory signals impinging on the brain from internal organs and from the outside. And, of course, memories strongly influence an emotional reaction. Indeed, the hippocampus is located just behind one of the key brain structures involved in emotional behaviors (the* amygdala*), and close links between the hippocampus and the amygdala have been proposed. These interactions are reciprocal; that is, we have more vivid memories of events that evoke a strong emotional reaction than of other events. Most of us over fifty can still remember exactly where we were and what we were doing when we heard that President Kennedy had been shot, an event that took place more than thirty years ago!*

The Autonomic Nervous System

The brain oversees two motor systems. One, the *voluntary motor system*, controls the muscles of the limbs, body, and head. It is the voluntary motor system that comes into play when we swing a golf club or swat a fly. The other motor system regulates our internal organs, including the heart, digestive tract, lungs, bladder, and blood vessels; this is the involuntary motor system or *autonomic nervous system*. As the name "autonomic nervous system" implies, control of the internal organs is mainly involuntary, which is why we are usually unaware of the effects of the autonomic nervous system on our internal organs. But this is not always the case. On occasion, it is possible to exert voluntary control over internal organs. Furthermore, there are important interactions that occur between the two motor systems. For example, when we voluntarily initiate strenuous activity like running or swimming, more blood flows to muscles critical for sustained activity, and this increased blood flow is mediated by the autonomic nervous system.

Two divisions of the autonomic nervous system exert opposing effects on most organs. One, the *sympathetic nervous system*, is known as the "fight or flight" system. It prepares us for action, and it is rapidly activated when we encounter a frightening or stressful situation.

Heart rate and heart output speed up, as does blood flow to the muscles. The pupils of the eye dilate to allow more light into the eye, and at the same time, digestive system activity slows down.

The second division, the *parasympathetic nervous system*, is the "rest and digest" system. When it comes into play, the body relaxes—heart rate and blood pressure drop, the digestive system becomes more active, and the pupils of the eyes constrict. After Thanksgiving dinner, the effects of the parasympathetic nervous system are obvious; we sink into a soft chair and doze off. Prior to dinner, during the predinner activities, the sympathetic system is in full swing and we are active and energetic.

Animals without a sympathetic nervous system can survive as long as they are maintained in a stable, warm, and comfortable environment. They cannot carry out strenuous activity or survive cold at all well. They also cannot cope with stress. When put in a stressful situation, they may even die, while animals with an intact sympathetic nervous system, when exposed to the same situation, cope perfectly well.

The anatomical organization of our autonomic nervous system, that is of our sympathetic and parasympathetic nervous systems, is special and distinct. In the sympathetic system, spinal cord neurons extend axons from the spinal cord that synapse upon (innervate) neurons in ganglia that lie along the spinal cord or in the abdominal cavity. The neurons in these sympathetic ganglia extend their axons to the organs they innervate, such as the heart, lungs, and digestive tract. In the parasympathetic system, the ganglia innervating an organ are located in the organ itself. The parasympathetic ganglia are innervated by neurons found in the brain stem or in the lowermost (sacral) part of the spinal cord. Many of the parasympathetic system axons coming from the brain stem are in two cranial nerves (one is the vagus nerve) that exit from the base of the brain and extend branches to various internal organs. The general organization of the autonomic nervous system is shown in *Figure 58*.

The sympathetic and parasympathetic nervous system ganglia are more than relay stations. The synaptic interactions that occur in them are complex and not well understood. Neurons and axons containing neuropeptides and other neuromodulatory substances are located in these ganglia, which means that subtle modulatory interactions happen within the ganglia. It is well established, though, that the neurons of the sympathetic ganglia release norepinephrine at their terminals, whereas acetylcholine is released from the terminals of the neurons of the parasympathetic ganglia. Thus, one can regulate the internal organs' levels of sympathetic or parasympathetic activity by giving patients blockers of one or the other of these agents.

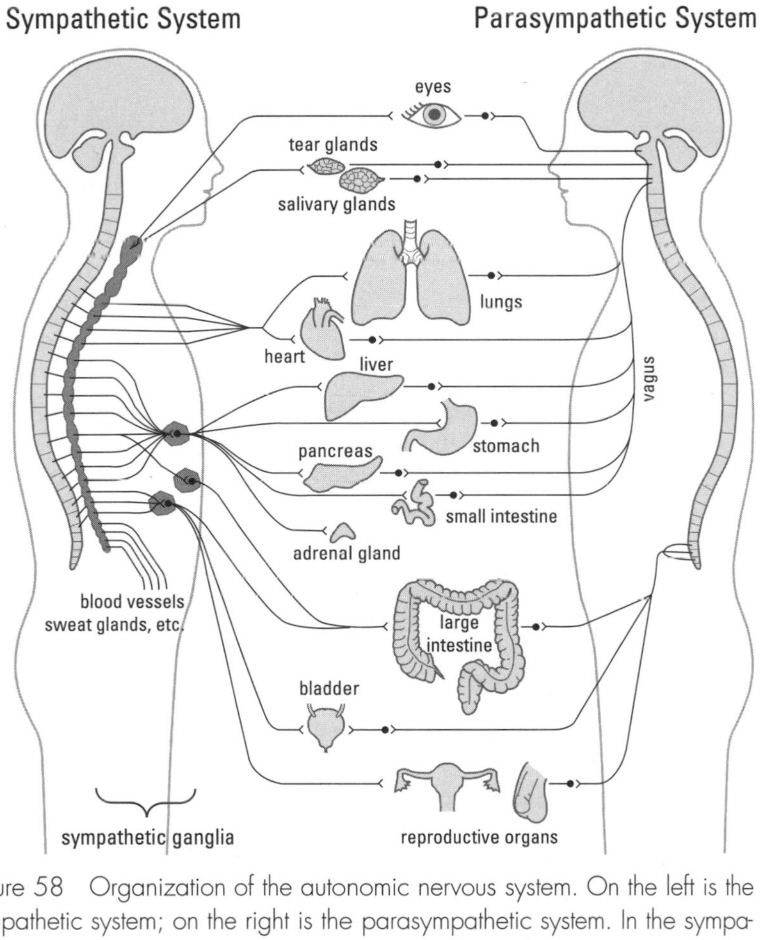

Figure 58 Organization of the autonomic nervous system. On the left is the sympathetic system; on the right is the parasympathetic system. In the sympathetic system, spinal cord neurons innervate a series of ganglia that lie next to the spinal cord or three ganglia found in the abdominal cavity. Neurons from these ganglia innervate the internal organs. In the parasympathetic system, motor neurons from the brain stem and sacral region of the brain stem innervate ganglia found in the internal organs themselves. Neurons in these ganglia innervate the host organ. The parasympathetic axons from the brain stem extend to the internal organs in various cranial nerves, particularly the vagus nerve.

How the two autonomic systems regulate an internal organ is elegantly illustrated by their effects on the heart (*Figure 59*). Heart cells have two types of neuromodulatory receptors. One is specific for the norepinephrine released by the terminals of the sympathetic neurons; the other is specific for the acetylcholine released by the parasympathetic nerve terminals. Both receptors are linked to G-proteins, which in turn are linked to the enzyme adenylate cyclase. (Recall that adenylate cyclase converts ATP to cyclic AMP and that cyclic AMP acts as a second messenger in cells and activates protein kinase A, or PKA; see Chapter 4.) However, norepinephrine activates adenylate cyclase via an *excitatory* G-protein, whereas acetylcholine inhibits adenylate cyclase activity via an *inhibitory* G-protein. Thus, sympathetic stimula-

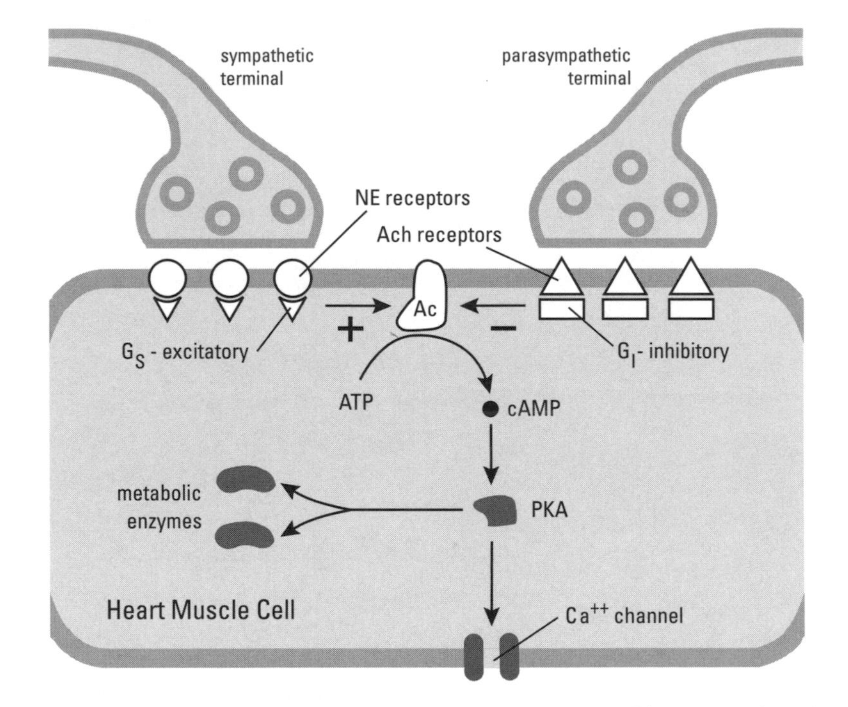

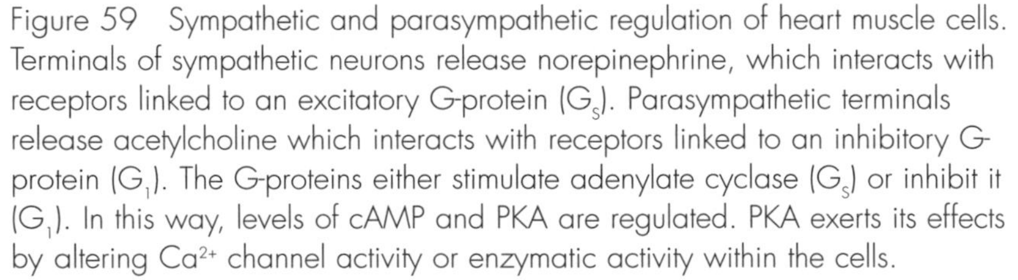
Figure 59 Sympathetic and parasympathetic regulation of heart muscle cells. Terminals of sympathetic neurons release norepinephrine, which interacts with receptors linked to an excitatory G-protein (G_S). Parasympathetic terminals release acetylcholine which interacts with receptors linked to an inhibitory G-protein (G_I). The G-proteins either stimulate adenylate cyclase (G_S) or inhibit it (G_I). In this way, levels of cAMP and PKA are regulated. PKA exerts its effects by altering Ca^{2+} channel activity or enzymatic activity within the cells.

tion increases cyclic AMP levels and PKA activity, whereas parasympathetic stimulation inhibits cyclic AMP production and PKA activity.

It turns out that PKA interacts mainly with Ca^{2+} channels in the heart cell membrane and it phosphorylates them. Phosphorylation of the channels allows more Ca^{2+} to enter the heart cell; decreased phosphorylation of the channels reduces the amount of Ca^{2+} entering the cells. Ca^{2+} exerts several effects on heart cells; it speeds up the heart rate and makes the heart muscle contract more strongly. It also raises the metabolism of the heart cells and induces the heart to beat more strenuously. Thus, the ultimate effect of sympathetic and parasympathetic stimulation on the heart is to regulate intracellular Ca^{2+} levels—increased Ca^{2+} levels stimulate heart activity, decreased Ca^{2+} levels depress heart activity. Thus we understand regulation of the heart by the autonomic nervous system down to the molecular and ionic level.

Hypothalamus

The autonomic nervous system is controlled by neurons found in nuclei of the hypothalamus. Hypothalamic nuclei regulate heart rate, blood pressure, respiration, gastrointestinal motility, temperature, and so forth, mainly through the sympathetic and parasympathetic systems. The hypothalamus also regulates the release of hormones from the pituitary gland, which sits just below the hypothalamus in the brain (*see Figures 30 and 61*). Certain of the hypothalamic neurons release small peptides into blood vessels that run between the hypothalamus and the pituitary gland. These blood-borne peptides induce (or inhibit) the release of hormones from cells of the pituitary gland. The pituitary gland hormones enter the general circulation and affect distant cells, tissues, or glands. Hormones released from the pituitary gland affect, for example, the adrenal, mammary, and thyroid glands, the ovaries, and the testes. Control of the endocrine system, especially the reproductive system, is another critical regulatory role of the hypothalamus. During a prolonged emotional or stressful situation, significant changes in endocrine function can result because of the effects of other brain centers on the hypothalamus.

During stress, for example, there is an increased secretion of a small protein, corticotrophin-releasing factor (CRF) from the hypothalamus. CRF causes the release of a pituitary hormone (ACTH), which travels via the bloodstream to the adrenal glands, which sit just on top of the kidneys. In response to ACTH, the adrenals release cortisol, a steroid hormone. Over the short term, cortisol can help in a stressful situation—it mobilizes energy stores and delivers them to muscles. But, if cortisol is continually released in response to a prolonged emotional or stressful situation, harmful results can occur. First of all, cortisol increases appetite, and this leads often to excessive weight gain during stress. But much more serious disorders have also been linked to high levels of cortisol in the bloodstream, including gastric ulcers, colitis, high blood pressure, impotency, and even excessive loss of neurons in the brain. It has even been suggested that prolonged stress can lead to premature aging of the brain and other organs. The immune system is also responsive to cortisol and is depressed when even moderately raised cortisol levels are circulating in the bloodstream. Thus, decreased resistance to disease has been linked to prolonged stress, and in animal experiments, an increase in tumor growth has been observed during stress. Blood levels of other hormones are also altered by stress or emotional states, and it is likely that harmful effects can result from these changes as well.

Emotional and Reinforcing Behaviors

Electrically stimulating the hypothalamus in animals evokes virtually every possible autonomic reaction, as was first shown in the 1930s by Stephen Ransom, an American neurophysiologist. He also discovered that stimulation of certain hypothalamic regions of animals evoked a panoply of emotional behaviors. Stimulating one hypothalamic region might induce rage or extreme aggressiveness in an animal, whereas stimulating another hypothalamic region might induce placidity in an otherwise wild animal. Stimulation of certain regions of the hypothalamus resulted in what appeared to be a pleasurable experience for the animal. If the experiment was set up so an animal could self-stimulate that hypothalamic region, it would do so continuously, to the exclusion of all other behaviors (*Figure 60*). These regions were originally termed *pleasure centers* and evoked much interest among neuroscientists. Although some brain regions outside of the hypothalamus will induce similar responses—animals will self-stimulate when any of these regions has electrodes implanted in it—the best and most reliable region to elicit this behavior is in the lateral part of the hypothalamus where a prominent bundle of axons extends to the forebrain.

Some insight into the significance of the pleasure centers has come from more recent experiments in which antagonists to dopamine were administered to animals, resulting in cessation of the self-stimulation behavior. Earlier experiments had shown that electrical stimulation of the lateral region of the hypothalamus releases dopamine in parts of the brain, particularly in a nucleus located in the basal region of the forebrain called the *nucleus accumbens*, a brain region involved in reinforcing behaviors such as drinking when thirsty, eating when hungry, or satisfying sexual desire. These behaviors have favorable consequences, and so they're repeated, which is why they are called reinforcing or rewarding. During reinforcing behaviors, dopamine is

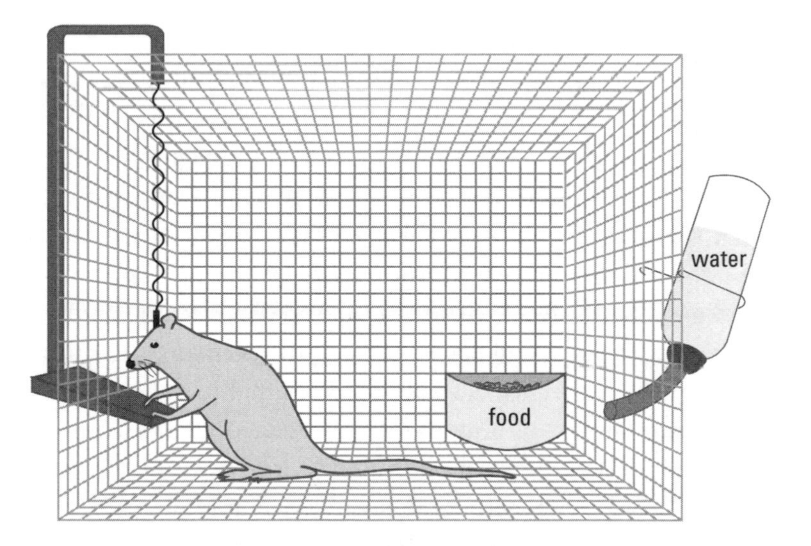

Figure 60 Experimental setup for electrical self-stimulation of the brain in the rat. When the lever is depressed, a small electrical pulse is delivered to the brain via an implanted electrode. Animals will do this to the exclusion of eating and drinking.

released in the nucleus accumbens, and agents that increase dopamine levels in the brain, including the amphetamines and cocaine, promote these reinforcing behaviors and mimic the effects of electrical stimulation of the lateral hypothalamus. Indeed, laboratory animals will press a lever to self-administer amphetamines or dopamine directly into the nucleus accumbens, much as they will press a lever to electrically self-stimulate the lateral hypothalamus.

The pleasurable and addicting nature of the amphetamines and cocaine are thus believed to be mediated by this reinforcing or rewarding brain system: The release of dopamine in the nucleus accumbens leads somehow to the pleasurable feelings. Other addicting drugs, including the opiates, marijuana, caffeine, and nicotine, also promote the release of dopamine in the nucleus accumbens; again, it seems likely that at least some of the pleasurable and addictive effects of these substances relate to this reinforcing or reward system.

The Amygdala, the Orbitofrontal Cortex, and Rationality

A subcortical brain region, the amygdala, which is located just in front of the hippocampus, has been implicated in integrating and coordinating emotional behaviors and in interacting with a number of key brain regions involved in emotional behaviors (*see Figure 61*). The first evidence of this surfaced in the late 1930s when investigators discovered that removing part of a monkey's brain that included the amygdala created a dramatic behavioral change in the animal. Very wild monkeys grew tame after the operation. And they became highly oral, putting virtually everything placed in front of them into their mouths, including repellent things like snakes. The animals also became hypersexual. The key structure underlying this behavioral syndrome is the amygdala, and lesions limited to the amygdala induce tameness and heighten oral and sexual activity in a variety of animals.

When a human's amygdala is electrically stimulated, the subject feels fear and anxiety; stimulating the amygdala of animals leads to a host of autonomic responses and emotional behaviors. The effects of stimulating the amygdala often mimic the effects of stimulating parts

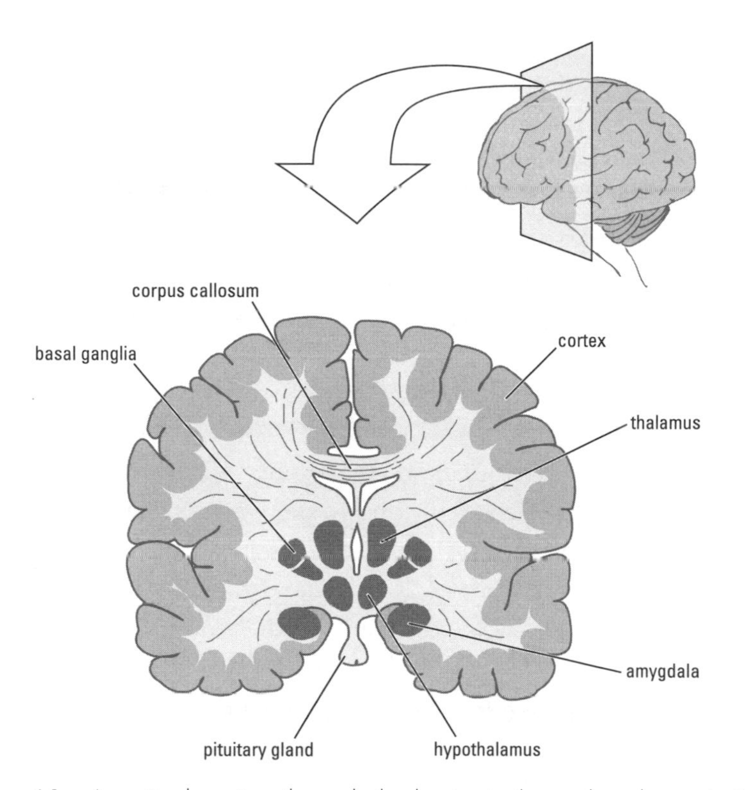

Figure 61 A vertical section through the brain similar to that shown in Figure 32, but a bit more anterior to show the position of the amygdala. The amygdala sits just in front of the hippocampus on the medial side of the temporal lobe.

of the hypothalamus. Indeed, the amygdala and the hypothalamus interact extensively, and many emotional and behavioral alterations that happen when the amygdala is stimulated are probably mediated by the hypothalamus and the autonomic nervous system.

The amygdala receives direct sensory input from the thalamus as well as input from the primary sensory areas of the cortex. The direct projection of sensory information from the thalamus to the amygdala is critical for more primitive emotional responses such as fright, whereas information from the cortex probably evokes more subtle and sophisticated emotional responses such as anxiety. Having direct input from the thalamus to the amygdala means that the amygdala—and subsequently the hypothalamus and autonomic nervous system—can be activated rapidly by sensory input, which is obviously advantageous in dangerous situations.

The amygdala thus plays a number of roles in emotional responses and behaviors. For example, various learned emotional reactions such as fear and anxiety require the amygdala. If an animal is conditioned to associate a sound with an unpleasant stimulus such as a shock to the foot, the animal will exhibit anxiety or even fear when it hears the tone in the absence of the stimulus. Lesions in the amygdala will abolish this fear conditioning because they prevent memories of fearful or anxiety-provoking situations from exerting any effects. But the amygdala also participates in positive emotional reactions. Animals will naturally seek out places or situations that are rewarding—where the animal has enjoyed pleasant experiences of food, water, and sex—and avoid places or situations that are not pleasant or not reinforcing. Again, animals with lesions in their amygdalas do not make these choices.

In addition to interacting with the hypothalamus, the amygdala has strong ties with cortical areas in the frontal lobes, regions essential for the appropriate expression of emotions. One of these cortical regions, just above the orbits of the eye, is called the *orbitofrontal cortex* (*Figure 62*). Lesions in the orbitofrontal cortex dampen normal aggressiveness and emotional responses in animals. Electrical stimulation of this area, producing manifold autonomic responses, prove that it interacts reciprocally with the amygdala and through the amygdala with the hypothalamus. The discovery that lesions in this part of the brain can

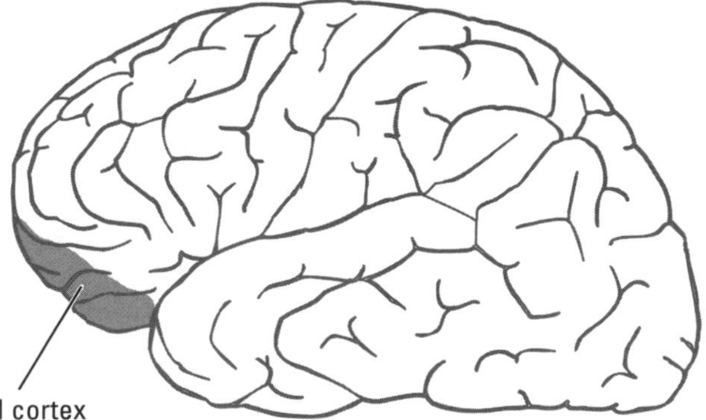

Figure 62 Surface view of the left hemisphere of the brain, showing the position of the orbitofrontal cortex.

induce a calming effect on aggressive monkeys, first reported at a scientific meeting in 1935, led to significant medical consequences for humans. A surgical procedure was developed for patients experiencing great emotional distress or extreme anxiety and was performed on thousands of patients, many in mental hospitals. This procedure, the prefrontal lobotomy, is largely discredited today after its widespread use in the 1940s and 1950s proved to be a major mistake, as we'll find.

The Frontal Lobes of Phineas Gage

Evidence that the frontal lobes play an important role in emotional behavior can be traced back to the well-known neurological case of a New England railroad construction worker named Phineas Gage. One day in the summer of 1848, Gage was packing dynamite into a hole in a rock with an iron rod when the dynamite exploded prematurely, shooting the rod into his cheek, through the frontal part of his brain, and out the top of his head. No one expected Gage to survive, but miraculously he did. Unfortunately, he changed into a very different man after the accident. Whereas before he was serious, industrious, and energetic, afterward he was irresponsible, irascible, and unpleasant. His self-concern was largely gone, as were his inhibitions. Emotionally he was a different individual whose personality was significantly altered.

John Harlow, the physician who treated him, wrote about the accident and his remarkable recovery, which was reported in Boston and Vermont newspapers. Harlow was well aware of Gage's personality changes and eventually reported them in the medical literature. Five years after Gage died, in 1861 in San Francisco, Harlow arranged to have Gage's skull shipped to the East Coast and to the Harvard Medical School, where it was kept on display in the Warren Medical Museum. (When I was a medical student there, I remember gazing with awe at the skull and the iron bar that had done the damage. Gage had kept the 13-pound, 3½-foot iron bar and it was buried with him. I was amazed that Gage could have survived such a severe brain injury.) From a modern analysis of Gage's skull we know that the site of brain damage after the explosion was the orbitofrontal cortex on both sides of the brain. Since then, several other medical cases with damage to the orbitofrontal cortex have been reported. Usually, these patients exhibit severe personality changes and altered emotional responses. The tumor that caused Elliot's problems, described at the beginning of this chapter, caused damage to the orbitofrontal cortex.

The 1935 experiment that led to the prefrontal lobotomy procedure involved two chimpanzees that would fly into rages whenever they were frustrated. The animals were involved in behavioral experiments, and whenever they made a mistake they displayed violent emotions. Two researchers, John Fulton and C. F. Jacobsen at Yale University, were trying to find out about the role of the frontal lobes in learning and recall. They removed one frontal lobe from the chimps to see what would happen to their performance on a recall task, and this did not change the chimps' behavior. Yet, when they removed the other frontal lobe, things were markedly different. No longer did the animals become frustrated and behave violently; instead they remained placid regardless of whether they made mistakes. They became much easier to handle and seemed much friendlier.

These observations were reported at the 1935 World Congress of Neurology in London. One of the listeners was a Portuguese neurologist, Egas Moniz, who wondered if removing or lesioning part of the frontal lobes in man might not relieve anxiety states and related psychoses. Another paper presented at that Congress reported that when a human patient had his frontal lobes removed, because of a massive tumor, he did not seem impaired intellectually—hence, it was concluded, humans could continue to function well without frontal lobes. Thus, Moniz, working with a neurosurgeon, began to lesion frontal lobes as a treatment for mental illnesses. The procedure did not actually remove the frontal lobes but severed the connections between the orbitofrontal cortex and the rest of the brain. The end result, however, was about the same.

There is no question that the procedure reduced anxiety, obsessions, and compulsions, and very disturbed and violent patients were often rendered more placid and tractable. Yet, many patients had severe side effects, however much they were generally ignored by those promoting the surgery. For example, patients given prefrontal lobotomies often became indifferent to the feelings of others and to the consequences of their own behavior. Whereas pathological emotional behaviors were often eliminated, normal ones were lost as well. The patients also had great difficulty making plans and managing their lives. The operation was too readily performed on inappropriate patients—patients who had a variety of mental disorders unrelated to emotional behavior—and so the surgery often debilitated them even further and was a disaster.

Today prefrontal lobotomies are seldom performed. Drug therapies are now quite effective in alleviating anxiety states, obsessions, and compulsions and in calming violent individuals. The one malady for which a modified prefrontal lobotomy procedure is sometimes carried out is intractable and excruciating pain. Following the surgery, such patients say that the pain is still present, but it doesn't seem to bother them any longer. The anxiety associated with the pain is relieved.

Antonio Damasio, a neurologist from the University of Iowa, has focused on the frequently made observation that individuals with frontal lobe lesions such as Phineas Gage or patients who have undergone a prefrontal lobotomy exhibit a dramatic change in personality. They often act irrationally, and have difficulty organizing their lives and planning ahead. Some become passive and dependent. And, as I have emphasized, subjects with frontal lobe lesions are disturbed emotionally—they do not respond appropriately to emotional situations, as the case of Elliot described by Damasio at the beginning of this chapter illustrates. Damasio links the loss of reason in these patients to their debilitated emotional state. He writes, "Certain aspects of the process of emotion and feelings are indispensable for rationality. Feelings point us in the proper direction, take us to the appropriate place in a decision-making space, where we may put the instruments of logic to good use."

Although at first glance this notion appears counterintuitive—we usually think of emotions as interfering with rational behavior—it is also the case, as Damasio points out, that strong feelings incite in us a plan of action. Without emotions and feelings, why bother? And this is the way patients with frontal lobe lesions behave. Elliot is typical of such patients. Not only do they respond passively to emotional situations affecting them, they also show a lack of concern about others. We think of reason and rationality as among the highest attributes of mind. Their close link to feelings provides a glimpse of the brain creating mind.

CHAPTER 10

The Conscious Brain

Human consciousness is just about the last surviving mystery. A mystery is a phenomenon that people don't know how to think about—yet. There have been other great mysteries: the mystery of the origin of the universe, the mystery of life and reproduction, the mystery of the design to be found in nature, the mysteries of time, space, and gravity. These were not just areas of scientific ignorance, but of utter bafflement and wonder. We do not yet have the final answers to any of the questions of cosmology and particle physics, molecular genetics and evolutionary theory, but we do know how to think about them. The mysteries haven't vanished, but they have been tamed. They no longer overwhelm our efforts to think about the phenomena, because now we know how to tell the misbegotten questions from the right questions, and even if we turn out to be dead wrong about some of the currently accepted answers, we know how to go about looking for better answers.

With consciousness, however, we are still in a terrible muddle. Consciousness stands alone today as a topic that often leaves even the most sophisticated thinkers tongue tied and confused. And, as with all the earlier mysteries, there are many who insist—and hope—that there will never be a demystification of consciousness.

—Excerpted from Daniel C. Dennett,
Consciousness Explained

No book on the mind can duck the burning question of consciousness. What is it? What underlies consciousness? We don't know the answers to these questions, but neuroscientists and others are thinking a great deal about these questions and attempting to relate consciousness to old and new findings on brain mechanisms.

We use the phrase "being conscious" quite ambiguously. A person who is asleep is termed unconscious, as is a patient in a coma. But being asleep and being in a coma are two very different states. When asleep, we are often mentally and neurally active, as when we dream. When a person is comatose, the brain is in a depressed state of activity. Comatose patients cannot be awakened even by intense sensory stimuli; usually they are totally unresponsive except for exhibiting basic reflexes such as the pupillary light reflex.

But even when one is awake and fully responsive, consciousness has different meanings. For example, much of our everyday activity is carried out "unconsciously." We brush our teeth, but usually are not conscious of the activities involved while doing it. We know we brushed our teeth this morning, but did we need to take the cap off the toothpaste tube (or was it already off?), and did we put the cap back on after we finished? (I just checked to see if I put the cap back on this morning—I did!)

"Awareness" is perhaps a better term to use for this kind of consciousness. At any one time we do many things that require mental activity, but we focus on one activity—what we are aware of at that moment. Later we can often recall and bring to awareness other activities happening while we were attending to something else. It is the bringing of something to consciousness or awareness that we want to understand. Can other mental states shed light on what might be going on when we are conscious in this restricted sense? Sleep and dreaming have been long thought to provide clues to mind and consciousness.

Sleep

Why we sleep is still a question we cannot answer. It is clear that we must sleep; indeed, when we are deprived of sleep we become extremely uncomfortable, can hallucinate, and can have psychotic reactions. Depriving animals of sleep can even lead to their death. Clearly sleeping is one of our strongest drives. We can starve ourselves, or refuse to drink, but it is virtually impossible to stay awake continuously for even a few days.

All mammals sleep, and probably all birds do, too. The evidence for sleep in most cold-blooded animals such as frogs and fishes is more equivocal, but all animals have periods of quiescence that may serve as sleep. It is also true, though, that mammals and birds have distinct stages of sleep that are not seen in cold-blooded vertebrates, so the equivalence of sleep behaviors between various kinds of animals is not clear.

An obvious reason for sleep is that it rests the brain and provides time for restoring neural processes. A corollary is that brain activity is diminished during sleep. But recordings from neurons during sleep indicate that they can be very active then. So sleep is now viewed as an active behavioral state, not an inactive one. It is true that some neurons and certain neural circuits are quiescent during sleep, but many neurons are as active during sleep as at other times, or possibly even more active. One current theory links sleep with long-term memory consolidation, and there is evidence that memories of certain tasks are improved following a period of sleep.

Distinct stages of sleep in mammals have been identified. Upon retiring for the night, we fall into a deep phase of sleep within thirty to forty-five minutes. During deep sleep, parasympathetic nervous system activity predominates; heart rate and blood pressure decline, but gastrointestinal motility increases. Muscles are relaxed, though we move every five to twenty minutes on average. After we have been asleep for about ninety minutes (or in deep sleep for about an hour), sleeping behavior changes. Sleep is shallower, and subjects can be much more easily roused by meaningful stimuli—calling their name, for instance. At the same time, muscle tone slackens and the limbs exhibit a partial paralysis. Even so, the eyes begin to move back and forth, sometimes very rapidly, as do the internal ear muscles. This stage of sleep, termed rapid eye movement or REM, is a highly active phase of sleep. Cerebral blood flow rises, as does brain oxygen consumption. And it is during REM sleep that most dreaming occurs and when some memory consolidation may happen. When subjects are awakened from REM sleep, most report that they have been dreaming; only a minority of subjects awakened from deep sleep say they have been dreaming.

REM sleep initially lasts for only twenty minutes, and then deep

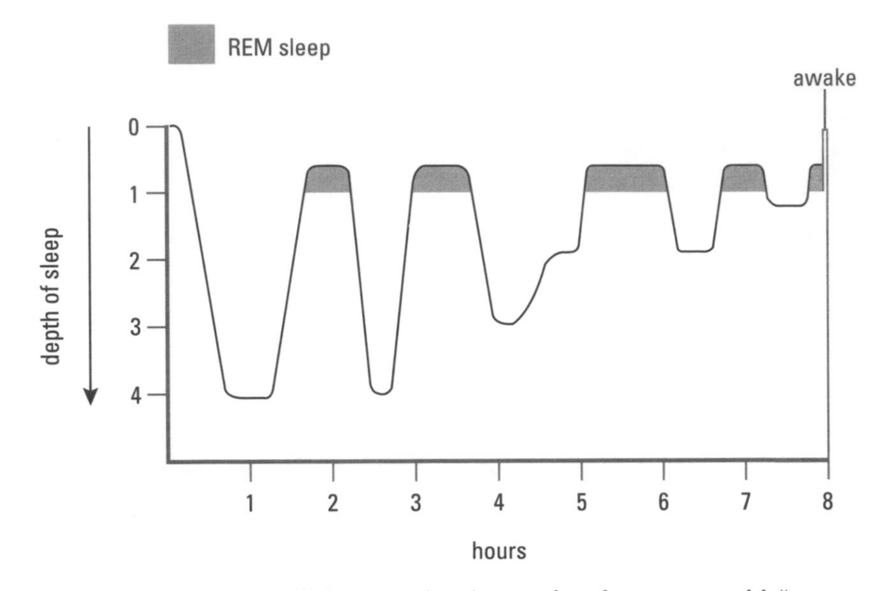

Figure 63 The stages of sleep. Within thirty to forty-five minutes of falling asleep, a deep phase of sleep occurs. This is interrupted after about an hour by a shallower phase of sleep and REM sleep. As the night progresses, the periods of deep sleep become progressively shorter and shallower whereas REM sleep becomes more prominent.

sleep ensues again for about ninety minutes. REM sleep recurs—for a longer period of time—and is followed by another phase of a deeper level of sleep. As the night wears on, the levels of deep sleep become shallower and shallower, and REM sleep periods typically become longer and longer, as shown schematically in *Figure 63*. Thus, over the course of the night, some four or five periods of REM sleep occur; so four or five periods of active sleep occur each night, and each active period is accompanied by dreaming. About one quarter of the night is spent in REM sleep, more of which is in the second half of the night.

Dreaming

The recognition of REM sleep and its correlation with dreaming has led to the realization that dreaming is a much more common phenomenon than was previously thought. Since ancient times dreaming has been thought important for understanding mental processes. Indeed, a famous book on dreams, written by Artemidorus of Daldis, dates to the second century A.D., and the dreams recounted in that volume are remarkably similar to modern ones. Yet it was generally assumed until recently that dreaming was a rare occurrence during the night. This is not so; we dream regularly during REM sleep, which means that each night we have many dreams, but most are forgotten—within eight to ten minutes or so after the cessation of REM sleep. We ordinarily remember dreams only from the late phases of sleep, from the time just before we wake up. Despite this, when subjects are awakened during REM sleep at any time during the night, at least three-quarters of the time they report that they have been dreaming.

Dreaming is clearly a distinctive state of mind. Most dreams—perhaps as many as two-thirds of them—are unpleasant and associated with apprehension, sadness, or even anger. Only a few are pleasant or exciting. As we all can attest, we are often frustrated in our dreams and subject to hostile acts. Dreams tend to be mainly visual in sighted people, but primarily auditory in blind individuals. Individuals who lose their sight gradually lose the ability to see in their dreams. Dreaming occurs also in deep sleep, but seems to be less common. Most nightmares that awaken children and adults in terror probably occur during deep sleep.

What role do dreams play? Much has been said about them but little has been concluded. Are they unique to man? All mammals and birds demonstrate REM sleep, and cats and dogs move during REM sleep as though they are dreaming. But whether they dream as we do is not known, just as it is not known whether their levels of consciousness or awareness are similar to ours.

Control of Sleep and Arousal

Lesions in the reticular formation of the medulla can make an animal stuporous or comatose. A comatose animal is totally unresponsive to external stimuli, whereas an animal in a stupor responds only to intense stimuli. Certain neurons in the reticular formation extend their axons widely throughout the brain, including the cerebral cortex, and are critical for maintaining the brain in an aroused state. Three principal neuroactive substances participate in this arousal system: norepinephrine, acetylcholine, and serotonin. Lesions of the reticular formation that destroy these aminergic and cholinergic neurons render an animal permanently comatose.

Neurons in the reticular formation also control sleep in an active fashion; that is, sleep occurs only when these neurons are activated. Furthermore, deep sleep and REM sleep are controlled separately. Certain neurons containing acetylcholine and located in the brain stem facilitate REM sleep. On the other hand, neurons in the brain stem containing norepinephrine and serotonin inhibit REM sleep.

How sleep is initiated is still not well understood. For many years it has been supposed that neurons in the brain release sleep-promoting substances. The search for such agents has been intense, and substances have been isolated, including two small peptides that will promote sleep in animals. But whether any of the substances so far isolated is a key sleep-promoting factor is not clear. Nevertheless, the enormous interest in discovering how sleep is initiated, controlled, and terminated, and by what factors, is undiminished, because insomnia is such a common and disruptive phenomenon. About 15 percent of people in industrialized countries have serious or chronic sleep problems, and another 20 percent have occasional insomnia. Furthermore, insomnia becomes more common among elderly people.

Sleep, like many biological processes, is partially regulated by a circadian rhythm or biological clock. Our internal clock, with a periodicity of about twenty-four hours, contributes to the regulation of sleep. The natural circadian rhythm that most of us possess is a little longer than twenty-four hours, but light each day resets the clock to approximately twenty-four hours. When humans (or other animals) are kept in constant darkness, the clock becomes free-running and for most animals and humans the circadian rhythm lengthens. This means that sleep comes later and later each day, as does awakening, so that the rhythm becomes out of synchrony with ordinary day and night.

How strongly circadian rhythms affect our normal sleep cycle is illustrated dramatically when we travel by air over several time zones, particularly to countries halfway around the globe. For many days after such a trip, most of us are awake much of the night and are very sleepy during the day. We gradually adjust to the new time zone, but it is not pleasant for the first few days, and we go through the same readjustment process when we return home.

Our body's main clock is located in a nucleus found in the hypothalamus. If this nucleus is destroyed in a rat, the animal's normal sleeping pattern is disrupted. The animal tends to sleep for short periods throughout the day and night, rather than mainly during the day (rats, being nocturnal, normally sleep during the day and are active at night). As is expected, this hypothalamic nucleus receives direct input from the retina, which is the pathway by which light resets the clock.

Consciousness and Awareness

What can we say about the neurobiology of consciousness? Throughout this book I have discussed how many brain phenomena can be linked with specific brain regions. The control of basic drives and acts occurs in the medulla and hypothalamus. Motor control and sensory processing can be related to a hierarchial series of cortical areas. Initial sensory processing occurs in the primary sensory areas and extends to association areas concerned with more specific aspects of sensory processing. The visual system, for example, has areas that primarily process color (V4) or motion (V5), and some specialized areas, such as one that recognizes faces.

Language can also be localized to certain cortical areas, as can different forms of memory. Even phenomena usually associated more with mind than brain, such as emotions and feelings, can be linked to specific brain structures, pathways, and regions. But what about consciousness? Whereas we can render an animal or human unconscious by lesioning the reticular formation in the brain stem, no one has provided any evidence that a region of the brain specifically relates to consciousness or awareness in the sense I have been defining it. This is not to say there is no such area; what I wish to suggest is that there may not be just one. Rather, consciousness may depend on many areas, working in concert.

Clearly our rich mental life depends on higher cortical function. The rapid development of the cortex during mammalian evolution made a big difference for humans, whose cortex is undoubtedly more developed than that of any other vertebrate or mammal. We possess mental abilities beyond those of any other organism, with language being the most obvious example. Humans can communicate ideas and images to one another either orally or by the written word, and this communication readily evokes images and emotions. We do not need to actually see something to visualize it—a poem or a narrative passage in a book can evoke intense images, sensations, and feelings.

Contrast this behavior with that of a frog, which "sees" quite well; it can visually identify a passing fly so that its tongue captures the fly when it is in front of the frog. But the fly must be moving for the frog to see it. A frog in a cage stocked with perfectly edible dead flies will starve to death; it simply doesn't see them if they are not moving. What, then, does the frog really see? Are its visual responses purely reflexive, much like our knee-jerk reflex? My guess is that they are: an animal without a cortex does not form visual images and does not see in the sense humans do. It responds to visual stimuli, but I would say unconsciously and without awareness. The same situation holds in human patients who have massive lesions in area V1, the primary visual cortex. Such patients have no awareness of seeing, yet they will exhibit visual reflexes. They may blink or even duck when a threatening object rapidly approaches them, but when asked why they blinked or ducked, they cannot explain why.

Our visual system constructs images based not only on incoming sensory information but also on experience. What we see depends very much on our expectations—what we have seen before—as well as on the information coming from our retinas. As noted earlier, vision is reconstructive and creative. The visual information coming from our eyes is imperfect; we construct a logical image based on incoming information and visual memories. If the visual information arriving by way of the optic nerve is ambiguous, we make a logical percept. But percepts can change, as the famous face-vase illusion beautifully illustrates. To make the percept change requires attention, much as focusing our consciousness on something requires attention.

In fact, what we see depends very much on memory and learning.

Indeed, we must learn to see. Many documented cases (including one described in Oliver Sacks's book *An Anthropologist on Mars*) tell of individuals who had been blind from birth or since early childhood, and whose vision was restored. The cataract or other impediment preventing their eyes from receiving appropriate visual stimuli was corrected. Yet, these individuals never learn to see, even though their "eyesight" is restored. They cannot effectively use visual information. They report seeing colors and vague objects, but they almost never become skilled visually. They continue to rely on other senses to function.

Learning to see occurs in the young human as does learning to speak. Just as it becomes more difficult to learn a language as one gets to be a teenager or older, it is difficult to learn to see when one is already an adult. And it is likely that this principle holds for other sensory modalities.

Consciousness, then, is a logical extension of the phenomena I have been examining: sensory perception, language, memory, and learning. When we focus our consciousness on something, we form mental images much as we do when we form visual images. We draw on memories to do this, but use the appropriate sensory systems to construct the images. Memories and images can evoke emotions and even bodily reactions by way of the autonomic nervous system. Language is a powerful stimulus for evoking such internal images. It may have been what first initiated our rich inner mental lives of consciousness. But now consciousness or awareness can be elicited entirely from within.

Is there any evidence for this view? Wilder Penfield observed that by electrically stimulating the temporal lobes, it was possible to evoke a memory so vivid and real that patients believed they were reliving an experience. Stimulating particular parts of the cortex evoked visual and auditory sensations along with emotions and feelings. Clearly, these experiences were evoked from within.

We know that animals can learn and remember things. What is it, then, about our memories that is special? Patient HM, who lost the ability to form long-term declarative memories when his hippocampi were removed, may provide a clue. He could still learn new skills and tasks, but he had no conscious memory of them. He learned things but had no awareness that he had learned them, and he even denied that he had done a learned task before. Is this what is special about our declarative memories—that when we elicit them we become aware of them? Are animals aware of what they learn? I don't think we really know.

The challenge before us is to understand these phenomena—sensory perception, memory and learning, language, and ultimately consciousness—in terms of neural mechanisms. At the moment this seems a daunting task, but we are getting glimpses of how neural activity can relate to complex behaviors. An elegant example has been provided by Patricia Goldman-Rakic and her colleagues at Yale University. Goldman-Rakic is interested in how humans and animals organize information so they can plan future actions. What is required is that incoming information, primarily sensory information, be combined with archival information—stored memories that are relevant. This process has been called *working memory*, and several lines of evidence suggest that such operations take place in the frontal lobes. (Recall from the last chapter that patients with prefrontal lobotomies typically experience difficulties in making plans and managing their lives.) Working memory, which is a form of short-term memory, is what we use to remember a phone number from the time we look it up to the moment we dial the number. Then, more often than not, we forget the number.

How neuronal activity participates in such phenomena is illustrated by recording neurons in the prefrontal area of the cortex—the most anterior part of the frontal lobes—while the animal performs a task requiring a delayed response. The test animals are monkeys and have permanent electrodes implanted in the prefrontal area. The monkeys soon ignore the recording electrodes and can be trained to perform a variety of behaviors. The task in question requires the monkey to "keep in mind" what it is supposed to do for a period of time after it has received the information needed to complete the task.

The monkey is first trained to look at a fixation point of light positioned in the center of its visual field. The animal is further trained to continue to look at this fixation point until it is turned off. While the fixation light is on, another light (a target) appears elsewhere in the visual field for a brief time. The monkey's task is to move its gaze to the location of the second light or target. But the monkey is trained to keep its eyes still until the initial fixation light goes off, which means the monkey must keep in mind where the target is for three to six seconds. If the animal does the task correctly, it is rewarded with a sip of grape juice.

What was found was that certain neurons in the prefrontal cortex become active during the time when the monkey is remembering where the target is. The neurons usually are inactive before the task begins, when the fixation light is on, and following the completion of the task. But they are active continuously during the delay between the turning off of the target and the fixation lights (*Figure 64*).

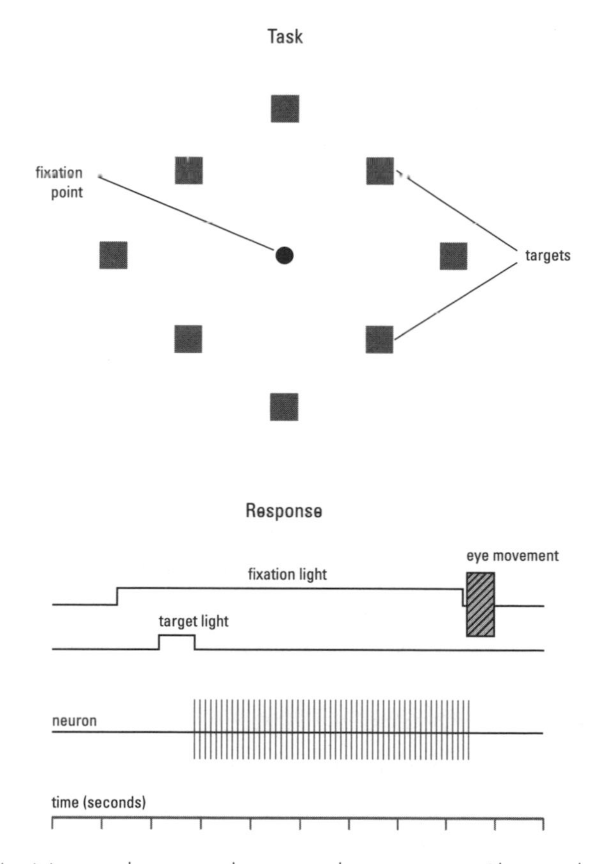

Figure 64 Neuronal activity during working memory. The monkey is trained to look at the fixation point. One of the target lights is then illuminated briefly. The monkey is taught not to move its eyes to the target until the fixation light is turned off, some three to six seconds after the target light is extinguished. Certain neurons in the frontal lobes become active and remain active during the period of time between the turning off of the target and fixation lights, or until the eye movement is made.

If the delay is excessively long, the neurons will gradually stop firing and the monkey will fail to perform the task. If during the normal delay period a monkey becomes distracted, the neurons will often stop firing, and again the animal cannot do the task reliably. (The same happens to us when we are distracted while trying to remember a phone number.) These neurons, then, are critical for the successful completion of the task. Their activity links the sensory stimulus and the eventual behavior.

An intriguing new twist has recently been added to the story. If the reward obtained upon the successful completion of the task is particularly appealing to the monkey—grape juice versus water, for example—the activity in certain prefrontal cortical neurons is higher. What this means is that the chances are now greater that the task will be successfully completed, suggesting how motivation can influence the neural activity underlying a behavior.

We are just at the beginning of understanding how neural activity underlies complex behavioral tasks. Few experiments such as the ones just described have been done, because they are so technically demanding. Furthermore, even the above example does not really tell us how neural activity keeps the location of the target in the monkey's mind. What is known is that a neuron codes for a specific location in the visual field. That is, some neurons will become active only when the target appears at the three-o'clock position in the visual field and others only when it appears at the twelve-o'clock position. Thus, the neurons seem to be coding for visual field position in addition to the delay.

The recording of neurons while animals, particularly monkeys, are awake and behaving will clearly tell us an enormous amount, and the techniques necessary to do this have now been quite well worked out. These are not easy experiments to do. The animals must not feel discomfort, so considerable effort is expended to ensure that they won't. It is impossible to carry out such experiments in humans; the neural basis of human consciousness seems beyond our experimental reach for the time being. Nevertheless, the noninvasive imaging techniques described in Chapter 8 promise to reveal much about neural function in humans. Maybe noninvasive techniques will enable us eventually to record the electrical activity of specific neurons, or at least groups of neurons, in the human brain.

The Future

Where will we be in fifty or a hundred years in terms of understanding brain mechanisms? No one knows, of course, but some speculations are possible. Certainly our increased understanding of brain mechanisms will have enormous impact on how humans conduct their affairs. Already this is the case. Two examples can be provided. That memories are often reconstructive, depending on past and present events and experiences, clearly has legal implications. We now recognize that distortion of memories is a normal phenomenon. Memories are made logical and coherent by our brains—and they are affected by emotion and traumas. That eyewitnesses remember events quite differently is inevitable and does not mean that someone is being untruthful. The perception of an event varies between individuals, for similar reasons, so discovering the "truth" about an event is a challenge, and perhaps an impossibility in some cases.

The recognition that the placebo effect relieving pain can be explained by the release of endogenous opiate-like substances within the brain raises the intriguing possibility that there are a variety of chemicals contained within neurons that can profoundly affect and alter the way we sense things or feel. The range of effects such substances might exert is conjecture at present—but that there are receptors in the brain activated by marijuana suggests there is an endogenous molecule in the brain that interacts with these receptors. Under what conditions is this molecule released, and what are its effects? These are a few of the questions that undoubtedly will be answered in the next few years.

That the brain may contain such molecules as its own opiate-like substances is influencing our notions of how we test for the effectiveness of new drugs and of how psychotherapy works. When a subject is given a test drug and told that it may help him, is the effect subsequently noted due to the drug or to an endogenous molecule released in the brain? And is the ultimate effect of psychotherapy to induce or alter the release of certain neuroactive substances within the brain? Clearly, mind-body interactions and influences are viewed quite differently today than they were just a few years ago. And it is safe to predict that psychiatry, viewed often as a lagging discipline in medicine, will be the major beneficiary of advances in understanding brain mechanisms.

Is there a Rosetta Stone for neuroscience, which if deciphered would transform the field as did the elucidation of the structure of DNA for molecular biology? Not that we are aware, but no one imagined that understanding the structure of DNA would give us such marvelous insights into the way genetic material replicates and how protein structure is coded. A discovery in neuroscience tomorrow might provide as powerful an impact on our field and on our understanding of how the brain works—we simply don't know where to look.

What are some of the more practical advances we can expect by the year 2050? The armamentarium of drugs to deal with mental illness and upsets will be larger and much more effective as we dissect further the intricacies of synaptic transmission, and drug design becomes yet more sophisticated. Schizophrenia, depression, panic attacks, anxiety, and even lesser problems such as jet lag will be controlled more effectively pharmacologically than they are today. And not only will symptoms be treated but also underlying causes. More specific drugs, and combinations of agents, will be available, fashioned to the mental profile of an individual patient.

Regeneration of central nervous axons will be possible to induce. Spinal cord injuries will not be the curse they are today—paraplegia and quadriplegia will be a thing of the past, much as is polio today. Optic nerve lesions will be treatable, damage after strokes minimized, and recovery after a stroke enhanced.

Our children will be routinely exposed to experiences and environments that will enable them to reach their full potential more easily. Learning foreign languages will be simpler, for example. They will comprehend languages better, and speak more like a native. Learning disabilities will be better understood and ways of dealing with them available.

These are only a few examples. They are just a sample—and a conservative one at that—of the possibilities current research promises.

Glossary

acetylcholine Neurotransmitter released at the neuromuscular junction. Also released from certain synapses in the brain, where it can have either neurotransmitter or neuromodulatory effects, and from parasympathetic nervous system neurons.

action potentials The transient, all-or-nothing electrical signals that travel down axons carrying the output information of neurons.

adenosine triphosphate (ATP) An energy-rich molecule made in mitochondria that powers biochemical reactions in cells.

adenylate cyclase The enzyme that converts ATP to cyclic AMP.

amacrine cell One type of inner retinal neuron. Many amacrine cells are movement-sensitive.

amblyopia The loss of visual acuity because of visual (form) deprivation of an eye or a crossed eye in a young animal or human.

amygdala A region in the forebrain involved in integrating and coordinating emotional behaviors.

amino acid The molecules that when strung together form proteins.

Aplysia californica An invertebrate used especially for studies of elementary forms of learning and memory.

areas V1–5 Areas in the occipital cortex concerned with processing visual information.

association areas Regions of the cerebral cortex concerned with higher levels of processing.

association neurons Cells that mediate interactions between neurons.

autonomic nervous system That part of the nervous system that regulates our internal organs. Much of the regulation is involuntary and mediated by two opposing subdivisions, the sympathetic and parasympathetic systems.

autoreceptors Receptors found on synaptic terminals that are activated by the substances released by the terminals.

axon Thin cellular branch that extends from a neuron to contact another neuron or a muscle cell. Axons carry the output message from the neuron via action potentials.

axon terminals Branches of an axon near its site of termination. Synapses are typically made by axon terminals.

axonal transport The special mechanism by which substances are moved rapidly down axons.

basal ganglia Five brain nuclei found in the forebrain that are concerned with the initiation and execution of movements.

bipolar cell A retinal neuron that carries the visual signal from the outer to inner retina.

bride of sevenless (boss) A mutation in the fruit fly that prevents the R7 photoreceptor from developing.

Broca's area An area (usually found in the left frontal lobe of the cerebral cortex) critical for the production of language.

calcium (Ca^{2+}) A positively charged ion important in synaptic transmission and that also serves as a second messenger.

calmodulin A protein that binds calcium ions. Activated calmodulin activates a specific kinase termed cam-kinase.

catecholamine A class of monoamine derived from the amino acid tyrosine. Dopamine, norepinephrine, and epinephrine are examples.

central nervous system Consists of the brain and spinal cord.

cerebellum A prominent hindbrain structure important for coordinating and integrating motor activity.

cerebral cortex A 2-mm-thick layer of cells that covers the forebrain. Highly infolded in man, the cortex is divided into two hemispheres, which are further subdivided into four lobes, frontal, parietal, occipital, and temporal.

cerebrum A collective term for the basal ganglia, cerebral cortex, and associated structures.

channel A membrane protein that allows ions to cross the cell membrane. Channels are usually closed until activated by a specific stimulus.

chloride (Cl^-) A negatively charged ion primarily involved in the inhibition of neurons.

circadian rhythm Endogenous rhythm that regulates various bodily functions depending on time of day.

complex receptive fields Receptive fields of neurons recorded in visual areas of the cerebral cortex that respond best to bars of light moving across the retina.

cone photoreceptors The photoreceptors responsible for color vision. Three types of cones exist in the human retina, respectively most sensitive to red, green, or blue light.

corpus callosum A thick band of axons found in the middle of the brain that carries information from one side of the brain to the other.

cortisol A steroid hormone released from the adrenal glands during stress.

cranial nerves Twelve nerves that enter the brain directly. Ten carry sensory and motor information related to the head, and two innervate the internal organs.

critical period The period of time during development when an animal is particularly sensitive to environmental conditions.

curare A drug that paralyzes muscles by blocking the acetylcholine receptors found on muscle cells.

current A measure of the flow of electrons through a wire, or ions across a membrane per unit of time.

cyclic AMP A second-messenger molecule formed by the enzyme adenylate cyclase from adenosine triphosphate (ATP).

cytoplasm The substance inside cells exclusive of the nucleus.

dark adaptation The time required for the eye to regain full sensitivity after light exposure.

declarative memory The memories of facts or events.

dendrites Bushy branch-like structures that extend from the cell body of a neuron and receive the synaptic input to the cell.

direction-selective cell A cell in the visual system that responds selectively to a spot or bar of light moving in a particular direction across the retina.

disinhibition The inhibition of an inhibiting neuron, resulting in a partial relief of inhibition.

dopamine A neuromodulator released from brain synapses that has been associated with two diseases, Parkinson's disease and schizophrenia.

eccentric cell Second-order neuron in the horseshoe crab eye.

ectoderm Cells on the outer surface of the embryo that become skin.

electrical synapse A junction where ions flow directly from one cell to the next.

electrons The small negatively charged particles that surround the protons in an atom.

endoderm Cells lining the inside of an embryo that form the gut and other internal organs.

enkephalins Small peptides released from synapses in the brain that have opiate-like effects.

epilepsy Seizures caused by diseased or damaged cells in the brain.

excitatory synapse A synapse that excites a neuron or muscle cell.

forebrain The most distal part of the brain, consisting principally of the thalamus, hypothalamus, basal ganglia, and cerebral cortex.

fovea The central region of the eye that mediates high-acuity vision.

frontal lobe The most anterior portion of the cerebral cortex, concerned primarily with movement and smell.

functional magnetic resonance (fMRI) scanning A technique that measures increases in blood flow in a region of the brain.

γ-aminobutyric acid (GABA) An inhibitory neurotransmitter in the brain.

ganglia Groups of nerve cells that usually serve a particular function. Typically used to describe groups of neurons outside the central nervous system or in invertebrate nervous systems, but there are exceptions, such as the basal ganglia in the forebrain.

ganglion cells The third-order cells in the retina whose axons form the optic nerve; also, cells in a ganglion.

generator potential A collective term for excitatory synaptic and receptor potentials that lead to the generation of action potentials.

glia Supporting cells in the brain that help maintain neurons, regulate the environment, and form the myelin around axons.

glutamate An amino acid that serves as the major excitatory neurotransmitter in the brain.

glycine An amino acid that serves as an inhibitory neurotransmitter in the brain.

G-protein A protein activated by postsynaptic membrane receptors. Usually linked to an enzyme that makes a second-messenger molecule.

gray matter Those regions of the brain and spinal cord where neuronal cell bodies and dendrites are abundant.

growth cones The specialized end of a growing axon.

growth factors Small proteins important for cell growth, differentiation, and survival.

guidepost neurons Specialized cells found in the developing brain that guide axonal growth.

habituation The decrease in the strength of a response following repeated elicitations of the response.

hemisphere (cortical) Half of the cerebral cortex. The two cortical hemispheres are each subdivided into four lobes.

hindbrain The lowermost part of the brain, emerging from the spinal cord. Consists mainly of the medulla, pons, and cerebellum.

hippocampus A region of the brain found under the temporal lobes that plays an important role in establishing long-term memories.

histology The microscopic study of tissues.

horizontal cell An outer retinal cell that mediates lateral inhibition between photoreceptors and bipolar cells.

horseshoe crab An invertebrate especially useful for studies of visual mechanisms.

hypothalamus A forebrain region that contains nuclei concerned with basic acts and drives such as eating, drinking, and sexual activity. The hypothalamus also regulates the release of pituitary gland hormones and the autonomic nervous system, and it plays an important role in emotional behavior.

Huntington's disease An inherited disease of the basal ganglia that causes movement dysfunction.

indoleamine A class of monoamine derived from the amino acid tryptophan. Serotonin is an example.

induction theory The theory that cells in the developing embryo become a particular type of cell depending on signals they receive from nearby cells.

inhibitory synapse A synapse that inhibits neurons.

interneurons A general term for all neurons that are found between sensory neurons and motor neurons.

invertebrates Lower animals such as insects, crabs, and molluscs that do not have a back-bone.

ion An atom that is charged—that is, has gained an extra electron and is thus negatively charged, or has lost one or two electrons and is thus positively charged.

kinases Enzymes that add phosphate groups to proteins, thereby altering their function.

lateral geniculate nucleus The nucleus in the thalamus that receives input from the eye and transmits the visual signal to the cerebral cortex.

lateral inhibition The reciprocal inhibition of one neuron by another, best characterized in the horseshoe crab eye.

L-dopa A precursor of dopamine. Useful for treating Parkinson's disease.

Limulus polyphemus The horseshoe crab.

lineage theory The theory that cells in the embryo are directed by inheritance to become a particular type of cell.

long-term depression (LTD) A persistent decrease in synaptic potential amplitude induced in a neuron by a strong priming stimulus delivered to the neuron.

long-term memory Memories that last for long periods—weeks, months, or longer.

long-term potentiation (LTP) A persistent increase in synaptic potential amplitude induced in a neuron by a strong priming stimulus delivered to the neuron.

Mach bands Light and dark bands seen adjacent to dark or light borders respectively that serve to enhance edge detection.

medulla A hindbrain region that contains nuclei involved with vital body functions, including heart rate and respiration.

membrane (cell) The thin barrier surrounding a cell that keeps various substances in and other substances out. Consists of a lipid bilayer in which are embedded various kinds of proteins, including channels, pumps, enzymes, and receptors.

mesoderm Cells between the ectoderm and endoderm in the embryo that develop into muscle, bone, and heart cells. In the early embryo, mesodermal cells induce overlying ectodermal cells to become neural plate cells.

midbrain That part of the brain between the hindbrain and forebrain.

mitochondria Structures found in cells that provide the energy-rich molecules (i.e., adenosine triphosphate) that power the cell.

monoamine A type of substance released at synapses that functions mainly as a neuromodulator.

monosynaptic reflex A simple reflex circuit consisting of a sensory neuron impinging directly on a motor neuron.

motor output The activity of motor neurons that results in muscle movements.

multiple sclerosis A disease of the myelin that surrounds axons.

myasthenia gravis A disease of the neuromuscular junction.

myelin An insulating layer of membrane formed around axons by glial cells.

nerve growth factor (NGF) The first-identified and best-studied growth factor found in the brain.

neural crest Cells derived from the neural plate that form much of the peripheral nervous system.

neural plate Cells found on the dorsal side of an embryo that form the nervous system.

neural tube An early stage in the development of the brain. Formed by the infolding of neural plate cells.

neuromodulator Substance released at a synapse that causes biochemical changes in a neuron.

neurons Cells in the brain involved in the reception, integration, and transmission of signals.

neuropeptides Small proteins (peptides) released at synapses that act mainly as neuromodulators.

neurotransmitter Substance released at a synapse that causes fast electrical excitation or inhibition of a neuron.

night blindness Loss of visual sensitivity at night. Can be caused by a deficiency of vitamin A or certain inherited eye diseases.

nodes Those regions along an axon where the myelin is interrupted and action potentials are generated.

norepinephrine A catecholamine released at certain synapses in the brain and also from sympathetic nervous system neurons.

nucleic acid (DNA) The genetic material found in the nucleus of a cell that codes for the proteins made by the cell.

nucleus A cluster of neurons in the brain that generally serves a particular function; also, structure within cells that contains the genetic material (DNA).

nucleus accumbens A nucleus found in the basal region of the forebrain involved in reinforcing behaviors such as drinking when thirsty or eating when hungry. This nucleus has also been implicated in addictive behaviors.

occipital lobe The most posterior portion of the cerebral cortex, concerned with visual processing.

oligodendrocytes Glial cells in the brain and spinal cord (central nervous system) that form myelin.

ommatidium The photoreceptive unit found in the eyes of invertebrates.

OFF cell A neuron whose activity decreases in response to stimulation. Often at the cessation of the stimulus such cells become more active for a short period.

ON cell A neuron whose activity increases in response to stimulation.

ON-OFF cell A neuron whose activity increases at the onset of a stimulus and again at the cessation of stimulation.

orbitofrontal cortex An area found in the lower part of the frontal lobes, important for the expression of emotional behaviors.

parallel processing The simultaneous processing of information along separate pathways.

parasympathetic nervous system The division of the autonomic nervous system that promotes "rest and digest" behaviors.

parietal lobe That region of the cerebral cortex between the frontal and occipital lobes concerned primarily with somatosensory information processing.

Parkinson's disease A disease of the motor system caused by a deficiency of dopamine in the basal ganglia. Patients with the disease typically develop a tremor and have difficulty initiating movements.

peptide A small protein.

peripheral nervous system Parts of the nervous system outside of the brain and spinal cord.

phosphorylation The addition of a phosphate group to a protein. Serves to modify the properties of the protein.

pioneer axons Axons that form early in development and provide a path for other axons to follow.

pituitary gland A gland found at the base of the brain that releases a variety of hormones into the bloodstream.

placebo An inert substance that can cause physiological effects under certain circumstances.

pleasure centers Regions of the brain which when stimulated appear to give pleasure to an animal. These regions appear to be related to the reinforcing and reward systems of the brain.

pons A hindbrain structure that relays information from the cortex to the cerebellum.

positron emission tomography (PET) scanning A method for detecting increases in activity of a part of the brain.

postsynaptic Pertaining to structures or processes downstream of a synapse; e.g., postsynaptic neuron, postsynaptic potential.

postsynaptic membrane That region of a cell membrane specialized to receive synaptic input.

potassium (K^+) A positively charged ion primarily involved in establishing the resting potential.

potential Another term for voltage.

prefrontal lobotomy A surgical procedure that severs the connections between the orbitofrontal cortex and the rest of the brain.

premotor areas Regions in the frontal lobes thought to be important for the planning and programming of motor movements.

presynaptic Pertaining to structures upstream of a synapse; e.g., presynaptic neuron, presynaptic terminal.

presynaptic synapse A synapse made onto a synaptic terminal.

primary motor area The region of the cerebral cortex where fine movements are initiated. Found in the frontal lobes adjacent to the central sulcus.

primary sensory area Regions where sensory information is first processed in the cerebral cortex.

procedural memory The memory of a motor skill such as riding a bicycle.

proprioceptive information Sensory information from muscles, joints, and tendons of which we are not aware—it does not reach our consciousness.

protein A chain of amino acids folded in complex ways that enable the molecule to carry out its prescribed function.

protons The positively charged particles found in the center of an atom.

pump A membrane protein that moves ions across the membrane of a cell. Pumps require energy to function.

Purkinje cell A large neuron found in the cerebellum.

pyramidal cell A prominent neuron found in all areas of the cerebral cortex.

receptive field That area of the retina which when stimulated causes a retinal cell to alter its activity.

receptors Membrane proteins found in postsynaptic membranes that are usually linked to intracellular enzyme systems; also, cells that respond to specific sensory stimuli, e.g., photoreceptors.

receptor potential The voltage change elicited in a sensory cell or neuron following the presentation of a specific sensory stimulus to the cell or neuron.

reflex An involuntary motor response in response to a specific stimulus.

REM *(rapid eye movement) sleep* An active phase of sleep during which most dreaming occurs.

resting potential The voltage across a cell membrane in the absence of any stimulus to the cell.

reticular formation Neurons found throughout the medulla that extend widely in the brain and are important for regulating states of arousal and levels of consciousness.

retinal The aldehyde form of vitamin A that, when combined with a specific protein, forms a visual pigment molecule.

retinitis pigmentosa An inherited disease that causes a slow degeneration of the photoreceptors.

retinular cells Photoreceptor cells in invertebrates.

rhodopsin The visual pigment of rods.

ribosomes Particles found in cells that are responsible for making proteins.

rod photoreceptors The photoreceptors responsible for dim-light vision.

schizophrenia A severe mental disease characterized by thought and mood disorders, hallucinations, etc.

Schwann cells Glial cells of the peripheral nervous system that form the myelin around axons.

sclera The outer (white) covering of the eye.

second messenger A small molecule synthesized in a cell in response to a neuromodulator (the first messenger).

sensitization The increase in the strength of a response following the presentation of an adverse stimulus to an animal.

serotonin A substance released at synapses that most often acts as a neuromodulator. Decreased levels of serotonin in the brain have been linked to depression.

sevenless A mutation in the fruit fly that prevents the R7 photoreceptor from developing.

short-term memory The initial storage of memories that lasts for fifteen minutes or so. Short-term memories are labile and easily disrupted.

simple receptive fields Receptive fields of neurons recorded in the primary visual area of the cortex that respond best to oriented bars of light or edges projected onto the retina.

sodium (Na^+) A positively charged ion involved in the generation of action potentials and in the excitation of neurons and sensory cells.

somatosensory Pertaining to sensory information coming from the skin and deeper tissues of the limbs and trunk, such as touch, pressure, temperature, and pain.

squid An invertebrate (a mollusc) that has giant axons.

stellate cell An association-type neuron found in the cerebral cortex.

sulcus A prominent and deep infolding of the cerebral cortex.

supplementary motor area A premotor area involved in the planning and programming of motor movements.

sympathetic nervous system The division of the autonomic nervous system that mediates "fight or flight" reactions.

synapse The site of functional contact between two neurons or a neuron and muscle cell.

synaptic potential The voltage change produced in a neuron following the activation of a synapse impinging on the cell.

synaptic terminal A site where a synapse is usually made.

synaptic vesicles Small vesicles found at synapses that contain the chemicals released at the synapse.

tectum A midbrain structure, especially prominent in nonmammalian species, which integrates sensory inputs and initiates motor outputs.

temporal lobe The lateral-most part of the cerebral cortex, concerned with hearing and memory.

thalamus A forebrain region that relays sensory information to the cerebral cortex.

topographic Pertaining to the orderly projection of axons from one region of the brain to another.

transcription factors Proteins that turn on and off the expression of genes by binding directly to the genetic material (DNA).

transporter A membrane pump that transports substances released at a synapse back into the synaptic terminal, thereby terminating the activity of the substances.

tricyclics Drugs that raise the levels of monoamines in the brain by inhibiting their reuptake into synaptic terminals.

tryptophan The amino acid from which indoleamines such as serotonin are derived.

tyrosine The amino acid from which catecholamines such as dopamine and norepinephrine are derived.

visual pigments The molecules in photoreceptors that absorb light and lead to the excitation of the cell.

voltage A measure of the electrical charge difference between two points; in neurons, the charge difference across the cell membrane (i.e., membrane voltage or potential).

voluntary motor system The part of the nervous system that controls the muscles of the limbs, body, and head. The control is mainly voluntary.

Wernicke's area An area in the left temporal lobe concerned with the comprehension of language and reading and writing.

white matter Regions of the brain and spinal cord where there are abundant myelinated axons. The myelin gives the tissue its whitish appearance.

working memory A memory maintained for a short time to enable a specific task to be accomplished. An example is remembering a phone number until it is dialed.

X-chromosome The so-called sex chromosome. Males have just one, but females have two.

Further Reading

A collection of books on various brain-related topics. All are accessible and most are a good read.

Barondes, Samuel H. *Molecules and Mental Illness*. New York: Scientific American Library, 1993.

Cajal, Ramón Y, S. *Recollections of My Life*. Cambridge, Mass.: M.I.T. Press, 1989.

Carlson, Neil R. *Physiology of Behavior*. Boston: Allyn and Bacon, 1994.

Crick, Francis. *The Astonishing Hypothesis: The Scientfic Search for the Soul*. New York: Charles Scribner's Sons, 1994.

Damasio, Antonio R. *Descartes' Error: Emotion, Reason, and the Human Brain*. New York: Grosset/Putnam, 1994.

Dennett, Daniel C. *Consciousness Explained*. Boston: Little, Brown, 1992.

Dowling, John E. *Neurons and Networks: An Introduction to Neuroscience*. Cambridge, Mass.: Belknap/Harvard University Press, 1992.

Hobson, J. Allan. *The Dreaming Brain*. New York: Basic Books, 1989.

Hubel, David H. *Eye, Brain, and Vision*. New York: W. H. Freeman, 1988.

Kandel, Eric R., Shwartz, James H. and Jessell, Thomas M. *Essentials of Neural Science and Behavior*. Norwalk, Conn.: Appleton & Lange, 1995.

Kramer, Peter D. *Listening to Prozac*. New York: Penguin USA, 1997.

Pinker, Steven. *The Language Instinct*. New York: Morrow, 1994.

Pollen, Daniel A. *Hannah's Heirs: The Quest for the Genetic Origins of Alzheimer's Disease*. New York: Oxford University Press, 1993.

Posner, Michael I. and Raichle, Marcus E. *Images of Mind*. New York: Scientific American Library, 1994.

Sacks, Oliver. *An Anthropologist on Mars*. New York: Alfred A. Knopf, 1995.

Sacks, Oliver. *The Man Who Mistook His Wife for a Hat and Other Clinical Tales*. New York: Harper Collins, 1987.

Schacter, D. L. *Searching for Memory: The Brain, the Mind and the Past*. New York: Basic Books, 1996.

Snyder, Solomon H. *Brainstorming: The Science and Politics of Opiate Research*. Cambridge, Mass.: Harvard University Press, 1989.

Valenstein, Elliot S. *Great and Desperate Cures: The Rise and Decline of Psychosurgery and Other Radical Treatments for Mental Illness*. New York: Basic Books, 1986.

Zeki, S. *A Vision of the Brain*. Oxford: Blackwell, 1993.

Index

Page numbers in *italics* refer to illustrations.

α-bungarotoxin, 37–38, 40
acetylcholine, 4–5, 44–45, 48, 96, 165–66, *166*, 182
ACTH, 168
action potentials, 25, *26*, 27–32, *28*, *30*, 31–32, 33, *34*, 44, 62–63, 65, 66, 70, 79
acupuncture, 57
adenosine triphosphate (ATP), 45, *46*, 78, *78*, 102, 166
adenylate cyclase, 45, *46*, 48, 78, *78*, 166, *166*
aging, 10–12, 168, 182
Aguayo, Albert, 84
A^- ions, 23, *24*
alcohol, 49–50
Ali, Muhammad, 12
Alzheimer's disease, 3–4, 11–12
amblyopia, 137, 139–40
amino acids, 48–50, *51*, *55*, *56*
amnesia, 75
amphetamines, 42, 50, 52, 170
amygdala, 163, 170–72, *171*
animals, 14, 20, 57, 185, 186–88
  cold-blooded vertebrates, 80, 84, 90, 91–92, *91*, 179, 184
  dreaming by, 181
  invertebrates, 30, 32, 61–80, 84
Anthropologist on Mars, An (Sacks), 101–2, 185
anxiety, 9, 49–50, 52, 59, 171, 172, 174–75, 190
aphasia, 145, 146, 147
Aplysia californica (sea snail), 62, 71–80, *73*, *74*, *76*, 153, 155
Artemidorus of Daldis, 181
aspartate, 48
atropine, 45
autoimmune diseases, 32, 39–40
autonomic nervous system, 93, 125, 162, 163–67, *165*, 171, 185
autoreceptors, *46*, 47
awareness, 178, 183–88
axonal transport, 65–66
axons, 6, 7–9, *7*, 67, 68, 82–89, 92, 169, 182
  action potentials generated by, 25, *26*, 27–32, *28*, *30*, 33, *34*, 44, 62–63, 65, 66, 70, 79
  development of, 131–36
  giant, of squid, 61, 62–66, *64*
  nodes along, 31–32, *31*, 33
  optic nerve, 103–4, *103*, 108, 110
  proprioceptive, 82
  regeneration of, 83–84, 190
  spinal cord, 85–89, *85*, *87*

barbiturates, 49–50
basal ganglia, 52, 90, 94–96, *95*
benzodiazepines, 49–50
biological clock, 182–83
biological warfare agents, 37
blind spot, 120–21, *121*
Bliss, Timothy, 153
Bloch's law, 68
blood-brain barrier, 52
blood flow, 163, 179
  in mental activity, 13, 156, *157*
botulism, 37
brain, architecture of, 81–99
Broca, Pierre Paul, 145
Broca's area, 145, 146–47, *146*, 149, 157, *157*

calcium (Ca^{2+}) ions, 35, *36*, 47, 49, 77, 79, 153, 154, *166*, 167
calmodulin, 47, 49, 77
cam-kinases, 47, 49
cataracts, 137, 139
catecholamines, 50

cell lineage theory, 128
central nervous system, 81–99, *85*
cerebellum, 7, 14, *15*, 90, *90*, 92, 93–94, 152
cerebral cortex, 5, 14, *15*, 90, *90*, 91, 92, 93–94, *95*, 96–99, *97*, 162–63, 171, 182
  consciousness and, 183–88
  electrical stimulation of, 149–51, 156, 168–71, *169*, 172, 185
  hemispheres of, 88, 90, 93, *95*, 96–97, 146–48, *146*
  lobes of, 96, 97–98, *97*
  orbitofrontal, 161–62, 172–76, *172*, 186–88
  organization of, 17–18
  representational maps in, 98–99, *98*, 150
  visual processing in, 17–18, 99, 101–2, 105, 113–18, 136–40, 141, 150, 183, 184–85
cerebrum, 90–91
channels, *21*, 22, 23, 25, 27, 31, 32, 33, 37–38, 39–40, 43, 44, 66
chloride (Cl^-) ions, 22, 23, 33, 44, 49, 50, 65
circadian rhythms, 182–83
Clostridium botulinum, 37
clozapine, 52, 54
cobra venom, 37–38
cocaine, 9, 35, 38, 170
Cole, Kenneth C., 65
color blindness, 101–2, 107–8
color vision, 99, 101–2, 105, 107–8, 110–11, 116, 119
coma, 19, 178, 181–82
complex cells, 114–16, *115*
cones, 105–8, *106*
consciousness, 84, 88, 92, 96, 162, 177–90
  awareness and, 178, 183–88
  sleep and, *see* sleep
corpus callosum, 90, *90*, *95*, 147, *148*
corticotrophin-releasing factor (CRF), 168
cortisol, 168
cranial nerves, 86, 92
critical period, 139, 140–41
curare, 37, 39, 45
cyclic AMP, 45–47, *46*, 48, 78, *78*, 153, 166–67, *166*
cysteine, 48, *55*

Damasio, Antonio R., 161–62, 175
dark adaptation, 106–7
dendrites, 7–9, *7*, 16, *26*, 29, 33, *34*, 86, 131
Dennett, Daniel C., 177
depression, 9, 41–43, 52, 53–54, 59, 96, 190
deprivation, visual, 136–40
development, 14, 123–42
  of axons, 131–36
  environmental factors in, 124–25, 128–30, 133–34, 136–40, 141–42
  language acquisition in, 136, 140–42
  neural plate in, 125–28, *126*
  neural tube in, 125, 126, *126*, *127*
  neuronal differentiation in, 128–31
  plasticity of, 135–36, 141–42
  theories of, 128
  of visual cortex, 136–40, 141
dialysis, 25, 40
diazepam (Valium), 9, 50
digestive system peptides, *55*, *56*
diplopia, 39
disinhibition, 70
DNA, 9, 130–31, 190
dopamine, 50, 51–53, *51*, 54, 56, 96, 169–70
dreaming, 178, 179, 180–81
drugs, 9, 35–38, 39, 41–59, 170, 189
drug therapy, 175, 189, 190
  antianxiety, 49–50, 59
  antidepressant, 42–43, 53–54, 59
  antipsychotic, 51–53, 54
  for pain, 57–58, 59
  side effects of, 54, 59

eccentric cells, 66, *67*, 68–70
effector cells, 25
electrical stimulation of brain, 149–51, 156, 168–71, *169*, 172, 185
emotional behavior, 93, 161–76, 183
  amygdala in, 163, 170–72, *171*
  autonomic nervous system and, 162, 163–67, *165*, 171
  hypothalamus in, 93, 167–70, 171, 172
  orbitofrontal cortex in, 161–62, 172–76, *172*
  rationality and, 173–76
  reinforcing behaviors in, 168–70
endocrine system, 167–68
enkephalins, *55*, 56–59
environmental factors in brain development, 124–25, 128–30, 133–34, 136–40, 141–42
enzymes, 35, 38, 44, 45–47, *46*, 53–54, 78, *78*, 129, 166–67, *166*
epilepsy, 147, 149–52, 155
epinephrine, 50
eserine, 38, 39
excitatory synapses, 25–27, 32, 33, *34*, 43–44, 48, *49*, 70
eye, 103–5, *103*, 137–40, *138*
  of fruit fly, 128–30
  of *Limulus*, 62, 66–71, *67*, *69*, 110, 128
  see also optic nerve; retina; visual processing

face recognition, 99, 117–18, 183
Fields, Howard, 57
fluoxetine (Prozac), 9, 35, 38, 42, 43, 54
forebrain, 89, 90, *90*, 91, 104, 125, 169
fruit fly, 128–30
Fulton, John, 174
functional magnetic resonance imaging (fMRI), 13, 156–59

Gage, Phineas, 173–74, 175
γ-aminobutyric acid (GABA), 48–50, *49*, 96
ganglia, 72, 75–77, 86, 164–66, *165*

basal, 52, 90, 94–96, *95*
ganglion cells (retinal), *103*, 104, 108–12, *109*, *111*, *112*, 113, 116, 131–33
generator potentials, 27
genes, 44, 46, *46*, 79, 96, 124, 129–31, 155, 190
mutated, 129–30, *130*
for visual pigments coding, 107, 108
genetic factors, 12, 95–96, 107–8
gill-withdrawal reflex, 62, 72–80, *73*, *74*, *76*, 153, 155
Gilman, Alfred, 45
glial cells, 6, 10
myelin of, 30–32, *31*, 33, 62–63, 83, 86
oligodendrocytes, 83–84
Schwann cells, 83–84, 125
glutamate, 48–50, *49*, 55
glycine, 48–50, *49*, 55
Goldman-Rakic, Patricia, 186
G-proteins, 45, *46*, 47, 48, 78, *78*, 166, *166*
gray matter, 86, *87*
growth cones, 133–35, *134*
growth factors, 131
growth hormone, 93
guidepost cells, 133–34

habituation, 71–80, *74*, *76*, 153, 155
haloperidol, 52, 54
handedness, 146–47, 148
Harlow, John, 173
Hartline, H. Keffer, 64, 66–70, 104
heart, 19, 25, 48, 166–67, *166*
heroin, 56
hindbrain, 89–90, *90*, 125
hippocampus, 80, *148*, 151–55, *154*, 163, 185
histamine, 48
Hodgkin, Alan, *65*
hormones, 93, 167–68
horseshoe crab (*Limulus polyphemus*), 62, 66–71, *67*, *69*, 110, 128
Hubel, David, 136, 137
Huntington's disease, 95–96
Huxley, Andrew, 65
hypothalamic peptides, 55, 56, 167
hypothalamus, 90, *90*, 93, 167–70, 171, 172
biological clock in, 183
pleasure centers in, 168–70

illusions, optical, 120, *121*, 184
imaging techniques, 13, 155–59, 188
immune system, 32, 39–40, 168
indoleamines, 50
induction theory, 128
inhibitory synapses, 25, 33, *34*, 43–44, 48, 49, *49*, 50, 68–70, 110
injury (brain), 10, 12, 13–14, 88, 92, 99, 101–2, 116, 162
recovery from, 13–14, 17–18, 83–84, 92, 190
see also lesions
insomnia, 182
invertebrates, 30, 32, 61–80, 84

ions, 20–27, *21*, *63*
calcium $Ca^{(2+)}$, 35, *36*, 47, 49, 77, 79, 153, 154, *166*, 167
chloride (Cl^-), 22, 23, 33, 44, 49, 50, 65
potassium (K^+), 19, 22, 23–25, *24*, 65, 78–79, *78*
sodium (Na^+), 22, 23, 25–27, 29, *30*, 32, 35, 43–44, 49, 65, 66
iproniazid, 53–54
isoniazid, 53–54

Jacobsen, C. F., 174
James, William, 162
Japan, 141
jet lag, 182–83, 190

Kandel, Eric, 71–72
Keller, Helen, 83
Kennedy, John F., 18, 92, 163
kidney disease, 19, 25
kinases, 45–47, *46*, 77, 78–79, *78*, 129–31, *130*, 154, 166–67, *166*
Kosslyn, Stephen, 159
Kramer, Peter D., 41–42

lactic acid, 13
language, 143–48, 183, 184, 185
acquisition of, 136, 140–42, 190
Language Instinct, The (Pinker), 143–45
lateral geniculate nucleus (LGN), 94, 113, *114*, 137–39
lateral inhibition, 68–71, *69*, 110–11, *111*
L-dopa, 52, 53
learning, 42, 44, 47, 62, 71–80, 84, 184–86, 190
conditioned, 172
of motor skills, 94, 152
lesions, 89, 116–17, 156, 162
in amygdala, 170, 172
in basal ganglia, 95
in Broca's area, 145, 147, 149
in cerebellum, 94
in hippocampus, 151–53
in hypothalamus, 93
in optic nerve, 190
in orbitofrontal cortex, 172–75
in reticular formation, 92, 181–82, 183
from strokes, 13–14, 88, 101–2, 116, 190
in visual cortex, 17–18, 99, 184
in Wernicke's area, 146
see also injury
lethal injection, death by, 19
Levi-Montalchini, Rita, 135
life span, 10–11, *11*
Limulus polyphemus (horseshoe crab), 62, 66–71, *67*, *69*, 110, 128
Listening to Prozac (Kramer), 41–42
Lives of a Cell, The (Thomas), 61–62
Livingstone, David, 58
Loewi, Otto, 48
Lomo, Terje, 153

long-term depression (LTD), 153–54
long-term potentiation (LTP), 153–55, *154*
LSD (lysergic acid diethylamide), 9, 42, 50

Mach, Ernst, 70–71
Mach band phenomenon, 66, 71, *71*, 110
Mangold, Hilde, 126–27
Man Who Mistook His Wife for a Hat, The (Sacks), 81–82
marijuana, 58, 170, 189
Marine Biological Laboratory (MBL) (Woods Hole, U.S.), 61–62, 63–65, 66
Marine Biological Station (Plymouth, England), 65
medulla, 90, *90*, 92–93, 94, 181–82, 183
membrane voltage, 22, 27, 29–30, *30*, 33, 43, 65, 79
memory, 11, 42, 44, 47, 49, 62, 71–80, 84, 96, 120, 122, 150–55, 163, 183, 184–88
  declarative, 152, 185–86
  distortion of, 189
  evoked by electrical stimulation, 150–51, 185
  formation of, 153–55
  long-term, 75, 79–80, 151–53, 155, 185–86
  older vs. newer, 155
  procedural, 152
  short-term, 75, 79–80, 152, 186
  sleep and, 179
  working, 186–88, *187*
midbrain, 89, 90, *90*, 91–92, *91*, 125
Milner, Brenda, 152
Missionary Travels (Livingstone), 58
mitochondria, 9
Moniz, Egas, 174
monoamine oxidase inhibitors, 53–54
monoamines, 48, 50–54, 56
monosynaptic reflex, 87
morphine, 56, 57
movement initiation, 84, 94–95, 96, 97
mudpuppy, 80
multiple sclerosis (MS), 32, 39
muscarine, 45
myasthenia gravis, 38–40
myelin, 30–32, *31*, 33, 62–63, 83, 86

naloxone, 57
neocortex, 18
nerve gases, 38
nerve growth factor (NGF), 135
neural crest, 125, *126*
neural plate, 125–28, *126*
neural tube, 125, 126, *126*, *127*
neuromodulators, 42–47, *46*, 48, 50–59, *51*, 78, 134–35, 153, 164–67
neuromuscular junction, 37–40, 44–45
neurons, 6–40
  differentiation of, 128–31
  electrical signaling by, 6, 20–34, 35, 43, 62–66
  long axon, 15–17, 16
  loss of, 10–14, 135, 168
  in nuclei, 16–18, *16*, *17*, 52, 92, 93, 94–96, 183
  number of, 6
  organization of, 14–18, *17*
  oxygen requirement of, 10, 12–13, 156
  plasticity of, 18
  short axon (association), 15–17, 16
  special nature of, 9–14
  tumors of, 10
  types of, 14–17, *15*
neuropeptides, 48, 55–59, *55*, 164, 167, 182
neurotransmitters, 27, 35–36, *36*, 37–38, 39–40, 43–50, 77–79, *78*, 134–35, 153
  acetylcholine, 44–45, 48, 96, 165–66, *166*, 182
  transporters of, 35, *36*, 38, *46*, 54
nicotine, 45, 170
night blindness, 107
nightmares, 181
norepinephrine, 50, 54, 165–66, *166*, 182
nucleus accumbens, 169–70

oligodendrocytes, 83–84
ommatidia, 66, *67*, 68–70, *69*, 129
opiates, 170
  endogenous, 56–59, 189
optic nerve, 84, 94, 103–4, *103*, 108, *114*, 120–21, 132–33, 184, 190
  of *Limulus*, 62, 66, 67–70, *67*, *69*, 110
orbitofrontal cortex, 161–62, 172–76, *172*, 186–88
oxytocin, 93

pain, 57–58, 59, 149, 175, 189
parasympathetic nervous system, 164–67, *165*, *166*, 179
Parkinson's disease, 12, 51–52, *53*, 95–96
Penfield, Wilder, 149–51, 156, 185
peripheral nervous system, 82, 84
Pert, Candace, 57
pesticides, 38
phosphates, organic, 38
phosphorylation, 45–47, 48, 77, 78–79, *78*, 129, 153, 167
photoreceptors, 45, 66–71, *67*, 102–3, *103*, 105–8, *106*, 110, *111*, 129–31, *130*
Pinker, Steven, 143–45
pioneer axons, 134
pituitary gland, *90*, 93, 167–68
pituitary peptides, *55*, 56
placebo effect, 57, 59, 189
plasticity, 18, 135–36, 141–42
pleasure centers, 168–70
poisons, 36–38, 39, 40
poliomyelitis, 88–89, 190
pons, 90, *90*, 92–93, 94
positron emission tomography (PET) scanning, 13, 156–59
postsynaptic membrane, 25–27, 35, *36*, 37,

39–40, 43–44, 45
potassium (K^+) ions, 19, 22, 23–25, *24*, 65, 78–79, *78*
prefrontal lobotomy, 173–75, 186
presynaptic synapses, 77, *78*
prizefighters, 12
proprioception, 81–83, 88, 94
protein kinase A (PKA), 45–47, 78, 166–67, *166*
Prozac (fluoxetine), 9, 35, 38, 42, 43, 54
psychiatry, 190
psychotherapy, 41–42, 59, 189
Purkinje, Jan, 14
Purkinje cells, 7, 14, *15*
pyramidal cells, 14, *15*

Ramón y Cajal, Santiago, 15–16, 123–24, 131, 133
Ransom, Stephen, 168
rationality, 173–76
Ratliff, Floyd, 68–70
receptor potentials, 25–27, *26*, *28*, 66
receptors, 44–45, 48, 50, 52, 54, 57–58, 78, *78*, 189
Reeves, Christopher, 83
reflexes, 84, 87, 89, 99, 178, 184
 gill withdrawal of *Aplysia*, 62, 72–80, *73*, *74*, *76*, 153, 155
reinforcing behaviors, 168–70
REM sleep, 179–81, *180*, 182
reserpine, 53–54
resting potentials, 22–25, *24*, 65
reticular formation, 92, 93, 94, 181–82, 183
retina, 8, 14, 80, 103–16, *103*, *109*, 119–21, 183, 184
 amacrine cells of, 111–12, *111*, *112*
 bipolar cells of, 108, 110, *111*, 119
 color of, 105
 cones in, 105–8, *106*
 ganglion cells of, *103*, 104, 108–12, *109*, *111*, *112*, 113, 116, 131–33
 horizontal cell of, 71, 110, *111*
 lateral inhibition in, 68–71, *69*, 110–11, *111*
 receptive fields of cells, 104, 108–12, *109*, *112*, 119, 136–37
 rods in, 105–8, *106*
retinal (vitamin A aldehyde), 106
retinitis pigmentosa, 108
rhodopsin, 105, 108
rhythmic movement, 84–85, 89
ribosomes, 9
Rodbell, Martin, 45
rods, 105–8, *106*
Roland, Per, 158
Rosenham, David L., 3–4
runner's high, 58

Sacks, Oliver, 81–82, 101–2, 185
sarin, 38
schizophrenia, 9, 42, 54, 190
 dopamine and, 51, 52–53, 54
Schwann, Theodore, 83
Schwann cells, 83–84, 125
Schwartz, James, 71–72
sea snail (*Aplysia californica*), 62, 71–80, *73*, *74*, *76*, 153, 155
Seligman, Martin E. P., 3–4
sensitization, 71–80, *74*, *76*, *78*, 153, 155
sensory stimuli, 20, 25–29, *26*, *28*, 181
serotonin, 50, *51*, 53–54, 56, 77–78, *78*, 182
Sherrington, Charles, 89
sign languages, 145
simple cells, 114, *115*, 116
size, perception of, 119, *120*
sleep, 178–83
 circadian rhythms in, 182–83
 control of, 181–83
 dreaming in, 178, 179, 180–81
 initiation of, 182
 stages of, 178–81, *180*, 182
Snyder, Solomon, 57
sodium (Na^+) ions, 22, 23, 24–27, 29, *30*, 32, 35, 43–44, 49, 65, 66
somatosensory information, 86, 97, 98
speech, 145, 146–47, 156–57, *187*
Spemann, Hans, 126–27
Sperry, Roger, 131–33
spinal cord, 6, 82–89, *85*, *87*, *90*, 92, 94, 125, 164, *166*
 degeneration of, 88–89
 function of, 84–85
 injuries of, 83–84, 88, 162, 190
squid, giant axons of, 61, 62–66, *64*
strokes, 13–14, 88, 101–2, 116, 190
Sullivan, Anne, 83
surgical procedures, 83, 147, 149–52, 155–56, 161, 185
 prefrontal lobotomy, 173–75
sympathetic nervous system, 163–67, *165*, *166*
synapses, *7*, 8–9, *8*, 16, 17, 18, 20, 22, 32–40, 42, 75, 77, 86, 134–35
 excitatory, 25–27, 32, 33, *34*, 43–44, 48, *49*, 70
 inhibitory, 25, 33, *34*, 43–44, 48, 49, *49*, 50, 68–70, 110
 mechanisms of, 35–36
 myasthenia gravis and, 38–40
 vulnerability of, 36–38
synaptic potentials, 25–29, *26*, *28*
synaptic vesicles, *8*, 9, 35, *36*, *37*, *46*, *79*
syphilis, 88
Szent-Gyorgi, Albert, 102

tectum, 91–92, *91*, 131–33, *132*
tetanus, 37
thalamus, 90, *90*, 94, 95, *95*, 113, 171
Thomas, Lewis, 61–62
thought disorders, 42, 52
time reference regression, 155
topographic projections, 131–33
transcription factors, 130–31, *130*

transplantation, cell, 52
transporters, 35, *36*, 38, *46*, 54
tricyclics, 54
tryptophan, 50, *51*, *55*
tumors, 10, 147, 161–62, 168, 174
tyrosine, 50, *51*, *55*

uremia, 19

Valium (diazepam), 9, 50
venom, cobra, 37–38
visual perception, 105, 118–22, 184
visual pigments, 105–8, *106*
visual processing, 101–22, 184
    in cerebral cortex, 17–18, 99, 101–2, 105, 113–18, 136–40, 141, 150, 183, 184–85
    learning of, 184–85
    in retina, *see* retina
vitamin A, 102, 106, 107
vitamin A aldehyde (retinal), 106
voluntary motor system, 163

Wald, George, 102, 107
Wernicke, Carl, 146
Wernicke's area, 146–47, *146*, 156, *157*
what pathway, 117, *117*, *157*
where pathway, 117, *117*
white matter, 86, *87*
Wiesel, Torsten, 136, 137
working memory, 186–88, *187*

X chromosome, 107

Young, John Z., *63–64*